The Robotics Revolution

With sincere gratitude for
friends', relatives', and numerous
colleagues' inestimable support

The Robotics Revolution

THE COMPLETE GUIDE FOR MANAGERS AND ENGINEERS

Peter B. Scott

Basil Blackwell

© Peter B. Scott, 1984

First published 1984

Basil Blackwell Publisher Ltd
108 Cowley Road, Oxford OX4 1JF, UK

Basil Blackwell Inc.
432 Park Avenue South, Suite 1505,
New York, NY 10016, USA

British Library Cataloguing in Publication Data

Scott, Peter B.
 The robotics revolution.
 1. Robots, Industrial
 I. Title
 629.8'92   TS191.8
 ISBN 0-631-13162-0

Typeset by Katerprint Co. Ltd, Oxford
Printed in Great Britain by
Bell & Bain Ltd, Glasgow

CONTENTS

Part III: Robotics in Action

Part IV: Social, Organisational and Economic Considerations

FOREWORD

Tom M. Husband

As the title of this book proclaims, we truly are in the midst of a robotics revolution. The speed at which both software and hardware have developed is breathtaking. Robotics applications are about to change the face of many work (and non-work!) activities.

In the early 1970s industrial engineers grudgingly conceded that the shop floor robot might eventually be useful in simple materials handling work. Soon the role perceived for the robot included straightforward spray painting and spot welding. By the late 1970s the momentum of the emerging technology forced even the most conservative and sceptical engineers to review the horizon. They now accepted that the second and third generation industrial robots available in the 1980s and 1990s would have a significant role to play in inspection, assembly and many other established activities long associated with human skills of manual and mental dexterity.

The picture is possibly even more dramatic in non-industrial applications. The use of robotics technology in space exploration, in security work, in farming, in helping the disabled and in a host of other fields of activity is quickly gaining acceptance.

From a strictly technological viewpoint this revolution is fascinating. Robotics engineers have developed an impenetrable jargon, robot-user clubs are mushrooming, hobbyist magazines are flourishing. Yet the impact of the robotics revolution is much too important to be left solely to the technologists. It is essential that others resist the jargon (or at least the worst of it) and set about broad understanding of the limits of this particular technological revolution.

There is a special need for social scientists, trade union officials and senior managers to get to grips with robotics. The effects of introducing robotics into an organization are not well understood. There is no empirical 'case law' yet available. It is clear, however, that many traditional work and organizational relationships will be altered. Equally, there are clear implications for skill displacement. From a strictly managerial perspective there is the question of assessing the cost-effectiveness of certain potential robotic applications.

For all of these reasons this book by Peter Scott is especially welcome. He tackles the subject on a broad front. The jargon is dealt with head-on. For readers who might flinch from such a confrontation he offers a convenient glossary at the end of the book. Questions of a social, economic and organisational nature are considered in an open and honest fashion.

It is often argued that robotics technology exemplifies all of the advanced technologies currently emerging in the developed world. Like information technology or biotechnology, robotics requires technical expertise on many fronts. It carries with it important social implications – relating particularly to employment levels; it implies major upheavals in the way we operate our organisations, the way we educate the younger members of our community, and re-train those in mid-career. It is a very persuasive argument.

Take the question of the breadth of technical expertise required. The robotics engineer needs a good understanding of mechanical, electrical, electronic, hydraulic and pneumatic engineering. He (or, increasingly, she) must also be conversant with appropriate software procedures. Yet we still do not train engineers in this fashion in the great majority of our universities. Robotics technology, like so many other technologies, call for *systems* engineers. Engineers in mid-career face particular difficulties in catching up, and keeping up, with developments across such broad frontiers.

This book is, I believe, the first to tackle these problems in the context of robotics. As the author points out, there are sections of the technical chapters which will seem simplistic to readers who are specialists in one particular field. Similarly, readers with a management background may be familiar with certain of the material on the economic aspects of robitics. It is very unlikely, however, that many readers will have been exposed to the breadth of material discussed in this book.

For this reason the book should prove particularly valuable for both student and professional coourses, in addition to providing an ideal introduction and overview for all busy professionals who may come into contact with robotics, in however slight a fashion. Final year undergraduates in, for example, computing will find much of value in the chapters relating to mechanical and control engineering. Similarly, undergraduate mechanical engineers will derive benefit from Peter Scott's treatment of software and sensor technology. The book will surely also be of great use to teachers and students on the growing number of specialist postgraduate courses in robotics and automation. For mid-career short courses the book offers a sound background text for a very wide range of mature students, whether managers or technologists.

At Imperial College we set up a Centre for Robotics and Automated Systems in 1981. We run postgraduate and short courses in robotics. We also pursue research and development activities involving robotic applications ranging from pork deboning to television assembly. In the course of our work we collaborate closely with engineers and managers from robot-maker and robot-user firms. We also supervise, naturally, young project students and research workers. In all of our dealings we meet a recurring problem. It is the need for a broad, basic, minimal awareness of the total technology of robotics. This includes an understanding of the economics and the social implications. Since 1981 we have regularly bemoaned the lack of a basic text which tackled the field in the necessary breadth.

I am delighted therefore that one of our own Centre colleagues has now produced the goods. I know this book will be invaluable to those we deal with at Imperial College. I am sure it will be equally useful to a very wide range of others in education, industry and elsewhere.

PREFACE

Of all the years to publish a book on robotics, 1984 must surely be the most theatrical! Yet the world we live in is, thank goodness, very different from that portrayed in Orwell's view of the future, and there is hope in that. I am essentially optimistic about the future of mankind and of the role robotics is to play in it, yet our options on both scores are far from settled.

As with any major discovery, robotics has potential for evil as well as good – indeed there are those who feel it is far from 'neutral'. Yet it must always be remembered that robotics is a *dynamic* discipline. It is, to a large extent, what we make it. Many have fears that robots may gradually take over the work of humans yet provide nothing to 'fill the vacuum'. But the field is not quite like a runaway train. If sufficient numbers wish to change its course it can be steered – as has been demonstrated by many government robotics initiatives throughout the world. Some roboticists, for example, suggest that enhancement of man (rather than his replacement as at present) is, in the long run to be preferred both economically and socially. Robotics will not decide on which course to take – people will. The responsibility for the robotic future is ours.

A decade ago I was studying science, and soon after, computing science. For fun, I learned everything that I could about robotics – even though I had been told that it was a 'a dead-end subject'! But times were changing, and in the nick of time great men like Joe Engelberger opened the world's eyes to robotics, so that I was able at last proudly to put ROBOTICIST on the sections of forms asking for 'occupation'! Suddenly I am living the science fiction of my youth.

It is pure luck to be paid to do one's hobby, but then to be asked to write a book on it, is a great privilege. I am extremely grateful to all those who have been involved at Blackwells, especially Tony Sweeney and René Olivieri who have helped me through the course. At Imperial College, many of my overworked colleages have gone beyond the call of duty and actually read and commented on some of my draft chapters, for which I give my sincere thanks – most of all to Tom Husband, friend and professor, without whose constant encouragement and practical support I could never have written the book. I am particularly indebted to him for being willing, despite his very heavy workload, to write a Foreword. Special mention must also go to Steve Bedley and all those others who so kindly sacrificed much time and peace of mind during the preparation of photographs and also to Francis Morgan, who provided such invaluable assistance during the preparation of the index. Finally, deepest thanks to my close friends and family who supported and helped me during the writing process – at the end of the day, however wonderful the robots, it is the *humans* who mean the most.

ACKNOWLEDGEMENTS

Photographs courtesy of:

ASEA Ltd: 4.7, 9.1, 16.2; S. Bedley, photographer, Imperial College: 1.10, 1.11, 2.1, 2.2, 6.1, 6.4, 7.3, 8.1, 8.2, 8.6, 13.1, 15.1, 16.1, 17.2b, 17.7; British Nuclear Fuels plc: 12.1; Laboratoire d'Automatique et d'Analyse des Systèmes, Centre National de la Recherche Scientifique, France: 17.2a; Industrial Robot Division, Cincinnati Milacron UK Co: 9.2, 10.4, 13.4, 17.4, 18.2; B. L. Davies, Imperial College, London: 1.2; The DeVilbiss Company Ltd: 6.2, 16.6; Engineering & Scientific Equipment Ltd: 1.1, 8.8; Fairey Systems, Fairey Automation Ltd: 7.2; 600 Fanuc Robotics Ltd: 3.4, 4.6; Ferranti plc, Dundee: 10.3; GEC Electrical Projects Ltd, Rugby: 6.3; General Electric Research and Development Centre (USA): 10.2, 12.5; Marconi Research Centre, GEC Research Laboratories: 1.9; Odetics Inc, Anaheim, California (USA): 12.6; Spine Robotics AB, Sweden: 13.2, 17.1; Wagner Indumat Systems Ltd: 12.2, 13.3, 18.3; Faculty of Science and Engineering, Waseda University, Japan: 12.4, 17.5; Zenith Data Systems Ltd: 12.3.

What's in it for you!
INTRODUCTION

Whether you are a manager, an engineer or anyone else who does not yet fully understand robotics (but wants to!) then do not worry – this book has been written for you. Naturally however, there are so many different facets to robotics, and so many different backgrounds to the people who wish to learn about them, that any author attempting to cater for a wide variety of readers (yet bring them all up to about the same 'level') has something of a problem on his hands.

As a result, this book has been written on the assumption that there will tend to be at least part of it which is slightly oversimplistic for a given individual. For instance, a mechanical engineer may already know much of the material in chapters 3 and 4, yet feel far less happy about the 'computing chapters' which follow. Similarly, a manager may be expert on economic justification (covered in chapter 16) yet may desperately want to understand how on earth robots actually work.

Whatever the reader's background, all chapters should be accessible if the book is read sequentially. The best part of 500 largely technical terms have been included in a glossary at the end of the book: not just robotic terms, but also related words which seem to come up in robotic conversations. For instance, 'fettling' is as common a term to mechanical engineers as 'polling' is to computing scientists and 'cash flow' is to managers, yet in each case 'never the twain shall meet'. The definitions in the glossary are of course the robotics-related meanings – words such as *cell* naturally have other meanings as well. Throughout the book, when a term included in the glossary is first mentioned (or mentioned again after a long gap) it is printed in **bold** type.

Although a very large number of 'real world' examples have been included in the text, detailed case-studies have been kept to a minimum for reasons of space. Interesting though they can be, each is only ever a *particular* way of doing things (and not necessarily the best) so, on the whole, the *general* approach which should be followed (applicable to the majority of cases) has been presented in their place.

Part I – Robotics background

This book has been split into five basic sections. Part I consists of two chapters which provide a background to introduce the subject and some of the basic concepts employed throughout the rest of the book. In chapter 1 the terms robot and robotics are explored, and the distant origins of the field are revealed – as far back as Ancient Greece men were building statues which moved as if they were alive. The difference between robotics and more conventional automation is they discussed, along with some of the closely allied mechanisms such as 'pick-and-place devices' and 'teleoperators'. In Japan, such devices are considered actually to *be* robots, so when trying to compare the numbers of 'robots' in different countries it is important to compare 'like with like'!

Nearly all industrial robots are really just robot *arms*, so these are considered in detail, and the different arm configurations are explained. Although some robot arms are quite similar in design to the human equivalent, others are very different – humans, for instance, do not have telescopic joints! Lastly, the differences between 'fixed' and 'variable sequence' robots, 'servo' and 'nonservo' robots, 'point-to-point' and 'continuous path' robots, and the different 'generations' are discussed, together with examples of just what they can do. As with any fast moving subject, jargon like this is bound to develop. Unfortunately, robotics has moved *so* quickly that even the definitions for some of the new terms are not universally agreed!

Chapter 2 considers the development of modern robotics from such unlikely beginnings as the mechanical calculating machines built in the seventeenth century and the automatic looms developed in France, through the early computers which could only manage a couple of instructions per second, up to the modern equivalents which can cope with millions. The developments leading to industrial robotics are next covered, from the embryonic stages during the industrial revolution (when even the idea of a 'factory' as a form of workshop was new to most people), to the explosion following the introduction of the Model T Ford (which could be completely assembled in only 90 minutes compared with twelve hours using traditional methods). Finally, robotics is considered worldwide, and the differences between countries are highlighted – are the Japanese really so much better than everybody else?

Part II – Robotics technology

This section contains the six 'technical chapters' covering the technology used in robotics. The reader really does not need to have a technical background to be able to understand these chapters, all of the technical terms are either explained or covered in the glossary. Nevertheless, the 'nontechnical' reader may find that he has to 'work at' these chapters slightly more than the others. Even so, it is well worth the effort to gain the broad understanding of just how a robot actually works.

Chapter 3 deals with the mechanical and mathematical aspects of robots, taking the reader through the techniques used to relate the positions of individual joints in a robot to the actual position of the robot gripper, as well as considering the special requirements of mobile robots. Chapter 4 is concerned with the various drive systems used for robots, which are usually electrical, hydraulic or pneumatic. The forms of transmission mechanisms commonly used with these drives are also covered, such as the clever 'harmonic drive' which allows a fast turning motor to be used efficiently to drive a slow moving robot joint. Finally, robot performance is considered and the problems about what is in fact meant by the 'accuracy' of a robot is tackled – does it for example refer to performance measured over a day or over a whole month?

Many robots are really controlled in the same way as the old steam engines! Just what this entails is covered in chapter 5. The concept of 'feedback' is of fundamental importance for accurate control, and this is explored in detail together with methods of applying it. Unless one's arm has gone numb one can 'feel' where it is even if one's eyes are shut. Just how a robot can sense the same thing is dealt with next, and the different devices it can employ are examined. Some assembly robots can sense their position to one hundredth of a millimetre, that is, a tenth of the thickness of a human hair! Next, the different forms of robot control strategy which can be adopted are explained, from simply moving from one point to another to sophisticated movement along a complex path.

How do you tell a robot what to do? This is explained in chapter 6. First, the sort of electronics involved are looked at, ranging from 'bubble memories' to devices for turning sensor signals into a form computers can understand. Then the two basic forms of teaching are considered in detail: a robot can either be given commands via a typewriter-style keyboard, or it can be physically moved through the sequence that is required. Nevertheless, there are a lot of programs inside a robot which are not written by the user, and these too are explained. Finally, examples are given of some of the existing robot languages, such as VAL and AML, and suggestions are made about the likely ways in which these will develop. One day it may be possible simply to issue the command, ASSEMBLE (PRODUCT 'A'), and then leave the robot with the task of determining just how best to do it!

The latest robots know what is going on around them. Just how they do this (and how *much* they can do) is explained in chapter 7. Wine tasting is not yet a requirement for robots, but vision, touch and even hearing and smell have already been employed industrially! In addition, robots can use senses which no human has: such as the same range-finding mechanism as bats. Of course, the sensors themselves are just part of the story– the signals they provide must subsequently be analysed and useful information obtained from them. The principles involved in such pattern recognition (for instance with vision) are explained towards the end of the chapter. Advanced developments of such vision work already allow an experimental system to be sensitive enough to respond appropriately to the expression on its inventor's face!

Part II itself finishes with a chapter discussing the different types of devices which can be attached to a robot (ranging from 'universal grippers' to welding torches), as well as the different methods of delivering objects so that robots can access them. Such mundane tasks as designing the feeders for components may not be as glamorous as building robots, but they are really just as vital when constructing a robotic system.

Part III – Robotics in action

With all the technical details of robotics out of the way, part III looks at the actual applications for which robots are *already* being used. The first of these four chapters looks at the wide range of tasks which robots can perform using nothing more than a gripper. These range from taking parts from a 'diecasting machine' to arranging or packing objects. Just how some of the industrial processes which have been robotised actually work is also explained. Chapter 10 looks at tasks which robots perform using a form of tool. These can involve applying paint, adhesives or sealants, welding or even cutting using water jets or lasers!

Industrial assembly can typically account for something like half of the total manufacturing cost of a product, and half of the required workforce. As a result, the possibility of using robots for the task is a very exciting one. Nevertheless, despite the adverts which refer to cars 'hand built by robots', in reality, there is still a long way to go before this is completely true – the vast majority of robots in the automotive industry are actually used for spot-welding. The whole of chapter 11 is concerned with robotic assembly, an application which is only just reaching industry. Not only are the various stages which must be followed in the assembly task very complex, but optimising the whole process can be a nightmare. Consequently, hints on how to tackle robotic assembly are also provided.

Finally in this section, chapter 12 deals with those existing devices which do not really come under the umbrella of 'industrial robots'. These include the remote controlled arms which are used for highly radioactive work, or in undersea or space exploration – none of which are really 'robots' at all! Nevertheless, many true mobile robots are being developed, and these range from vehicles such as forklift trucks adapted for robotic control, to robots which actually employ mechanical legs to walk with. The same sorts of techniques which are required to control such mechanical legs have in fact been applied to help human victims of paralysis haltingly to walk again!

Part IV – Social, organisational and economic considerations

This section is concerned with some of the nontechnical considerations of robotics, which are every bit as important as the more technical aspects. After all, if robots are unjustifiable on either economic or social grounds, then it does not really matter how technically superlative they are. Chapter 13 looks at the worries people have about the introductions of robots, and weighs up the pros and cons and social implications of the general trend towards robotisation. Choosing the right path is certainly not as clear cut as it might at first glance appear. For example, are the arguments that the current forms of robotisation are vital for economic survival really as watertight as they appear? Is there possibly a third alternative to 'automate or liquidate'?

The mere decision to employ robots carries no guarantee of success. The secret is to learn from the experiences of others, so in chapter 14 the problems of introducing robots into a factory are considered, complete with a detailed strategy which a company can follow in order to robotise (if appropriate) as painlessly as possible. What pointers are there for helping to decide whether a given company is not yet 'ready' for robotisation? How important is it likely to become to have technically conversant managers at the highest levels of a firm? How does the Japanese decision making process vary from that used elsewhere?

Robots can be killers! Chapter 15 considers the safety problems which have all too often been disregarded, occasionally with tragic consequences. There are many relatively simple solutions which can at least go a long way to protecting personnel (and machinery). The simplest is just to keep people away – but what happens when somebody needs to be beside the robot in order to program it? This chapter looks at the reliability of robots, and how different design techniques can be employed for robots destined for tasks such as unmanned space exploration, where a maintenance crew may find it a little difficult to correct any fault!

Chapter 16, the last chapter in this section, is concerned with the economic justification of robots. This is of vital importance, yet is typically totally mismanaged because we are not yet really used to dealing with machinery as flexible and as potentially integrated with other systems as robots. The far reaching implications of this tremendously important 'integration effect' are explained, and then, having specified and explained the wide variety of both direct and indirect factors involved in justifying robots, various approaches to investment appraisal are considered.

Simple 'payback' is not suitable for robotic evaluation (indeed it is debateable how appropriate it is for many of the investments for which it is automatically employed). The reasons for this are pointed out, and the detailed operation of more apposite methodologies, such as 'net present value' and 'internal rate of return', are explained. Although these two approaches are often thought of as slightly complicated, the principles behind them are really very straightforward. The increase in robot numbers makes it increasingly important to understand such techniques because using them one can obtain a far more accurate impression of the benefits of robotisation than with the more conventional techniques such as 'payback'. Lastly, this chapter contains some guidelines to follow when evaluating projects.

Part V – Current prospects

The two chapters in the final section of the book take the reader first behind the scenes of some of the research laboratories throughout the world, to see a cross-section of the work related to robotics being conducted there, and secondly into the unknown with predictions of the potential which robotics appears to have. Current robotics research ranges from flexible manufacturing systems (FMS) to robotic sheep shearing, from security guarding to deboning bacon backs and much more! There are a whole host of directions which robotics is following.

The final chapter considers the likely forms of robots throughout the 1980s and 1990s, and then considers the implications of the seemingly inexorable progression of the subject. Just at what point this chapter turns from science prediction into science fiction largely depends on you! Slowly, it is beginning to dawn on some of us that never before in his long history has mankind unlocked the secrets to a tool as potentially powerful as robotics. Within the remainder of this book you will begin to see why . . .

Part I
ROBOTICS BACKGROUND

Within only about five years, robotics has grown in the public eye from being science fiction to being the panacea for industry! Increasing numbers of people from a variety of backgrounds are becoming involved either through choice or through necessity. The two chapters in this section are concerned with firstly the 'basics', and secondly the historical background of the subject. They are a grounding for anyone who wishes to understand the Robotics Revolution.

1

What all the fuss is about

FUNDAMENTAL ROBOTICS

Robotics, not robots

This book is about the potentially very wide ranging field of 'robotics' as distinct from being solely about those 'robots' which happen to be currently available. To explain: the book has been written at a time of great excitement and energy within the areas of robot design and usage, and if it were devoted solely to the examples of robot technology available at the time of writing, then as you read this it would already be out of date.

The rate of development of much of robot technology is actually closely linked with developments in computing, and there is an often quoted statistic about computing – if the automotive industry had developed as fast, then a Rolls Royce of 30 years ago would now only cost £2, would run for its whole life on one pint of petrol, and would develop enough thrust to drive the QE2! This also provides some indication of just how rapidly robot technology is currently advancing. However, although the sophistication of the technology is racing forward at a staggering pace, nevertheless the principles which underpin this rapid development themselves change comparatively slowly. It is an understanding of these areas which is the real key to unlocking the secrets of the new and powerful robot explosion. This is the field of robotics.

Origins of robotics

The detailed evolution of modern robotics is dealt with in chapter 2, but man had been toying with the idea of somehow building a mechanical version of himself long before the tentative work was started which eventually led to the successful introduction of an industrial robot in the early sixties.

Throughout recorded history man has had a preoccupation with sentient machines made, at least partially, in his own image. In the Ancient

Greek myths, the God of fire, Hephaestus, was supposed to be accompanied and aided by two pure gold living statues, and he later constructed a brass giant, Talus, to guard Crete against all intruders. In real life, but still over two thousand years ago, Hero of Alexandria wrote his *Treatise on Pneumatics* in which he described a large number of **automata** such as moving figures and singing birds – rather like an Ancient Greek Disneyland. It is remarkable that such sophisticated toys were the only real use ever made of pneumatics for centuries.

About 1500, Leonardo da Vinci built a mechanical lion in honour of Louis XII which, when the king entered Milan, moved forward, clawed open its chest, and pointed at the coat of arms of France! Such mechanical automata remained the vogue for the next four centuries, with ever increasing complexity and sophistication, yet it was not until the early twentieth century that the actual word 'robot' came into the English language from the translation in 1923 of a Czech play RUR (*Rossum's Universal Robots*) by the philosopher Karel Capek (C pronounced as CH). The actual 'robots' in the play are in fact biologically grown and, other than a lack of emotions, are indistiguishable from humans. The term 'robot' itself is derived from the Czech 'robota' meaning 'compulsory labour' and from 'robotnik' meaning 'serf'. Although the creations in the play would now be termed 'androids' rather than robots (which are now considered to be mechanical) the misuse of the word is of course universal.

The word 'robotics' was invented by the master science fiction writer Isaac Asimov in his now classic robot stories which at the time of writing were unusual in that they told of robots which not only did not harm humans but actually helped them! It was in one of these stories called '*Runaround*', which appeared in the March 1942 issue of '*Astounding Science Fiction*', that Asimov first propounded the famous Three Laws of Robotics.

1 A robot may not injure a human being or, through inaction allow a human being to come to harm.
2 A robot must obey the orders given to it by human beings except where those orders would violate the First Law.
3 A robot must protect its own existence, except where that would violate the First or Second Law.

Although Asimov did not realise it at the time, this was the first printed appearance of the word 'robotics'. Joe Engelberger, founder of Unimation and considered by many to be the 'Father' of modern industrial robotics, has pointed out that the Three Laws remain worthy design standards for roboticists to this day.

What is a robot?

As can be imagined from the confused usage of the word 'robot' since RUR, conceptions vary of what a robot actually consists of. Even when it comes to the comparatively recent concept of the 'industrial robot' there is no international agreement about definitions – it is comparatively arbitrary where the 'boundaries' of the term are drawn. Thus, in Japan a **pick and place device** (a simple mechanical arm whose motions are governed by hitting mechanical end-stops – Figure 1.1) *is* termed a robot, whereas in the West, due to the lack of any inherent *flexibility* in the device (unless someone physically moves the end-stops) it is considered to be a special case of **fixed automation**, rather than a robot. Classification of different devices that are termed robots is just as fraught with problems and is dealt with later in this chapter.

The term 'automation' is in fact newer than the word 'robotics'! It was coined by John Diebold in a paper written at the Harvard Business School in 1950. The correct word at the time was actually 'automatization', but, claims Diebold, his spelling was so bad that he decided to

Figure 1.1 A typical (non-robotic) pick-and-place device

contract the word to make it easier, and the term caught on when he wrote a book of that title in 1952.

So when is something a robotic system and when is it merely a more conventional form of automation? Imagine the problem of cutting a piece of metal from a large sheet. We can conceive of all sorts of different levels of sophistication employed to accomplish the task, and because this is a book dealing with robotic systems rather than just robots we must consider both the cutting operation as well as the handling of the sheet. The various degrees of sophistication might be:

1. A man tries to bend the sheet back and forth with his bare hands until it fractures.
2. The sheet is cut using a hand tool.
3. The sheet is cut using a powered tool.
4. The sheet is cut using powered machinery totally under human control.
5. The cutting machinery automatically executes a predetermined cutting sequence which cannot be changed, and it is fed either by a human or a production line.
6. A pick-and-place device picks up a sheet from a single fixed position and drops it into the machine which then executes a predetermined cutting sequence. The position the arm picks up the sheet, and the actual cutting sequence, can both be changed if the machinery is physically adjusted.
7. A simple **point-to-point robot** picks up a sheet from one of many possible positions and then places it into the machine which cuts one of several possible profiles (possibly dependent on where the robot picked the sheet up from).
8. A **continuous-path robot** gently picks up one of many sheets and under controlled acceleration feeds it into the machine which proceeds to cut one of many complex shapes from the metal.
9. The whole robotic system is part of a much larger system all controlled by computer. The shapes cut might be determined by those parts required later on in the factory.
10. The whole robotic sytem makes substantial use of visual and tactile information, for example to find out where a sheet actually is.

Up to and including level 6 is considered (in the West) to be fixed (or dedicated) automation, although clearly 6 is becoming quite flexible! Level 7 is the simplest robotic system because the potential variability of programmed motions of the manipulator classifies it as a robot. The actual metal cutting machine would probably be a **computer numerical control** (CNC) device. Such an automatic machine tool is controlled by a mini- or microcomputer, using prerecorded details of some or all of the required machining sequence. However, although programmable it is not considered flexible enough to be termed a robot because a given CNC machine can only ever, for example, cut metal. Levels 9 and 10 have already found limited use on the factory floor but problems remain to be solved before they gain widespread acceptance.

As regards actual definitions for a robot, various ones have been adopted. The common usage of the word seems to include all those machines which wholly, or in part, imitate man – sometimes his appearance, sometimes his actions, and sometimes both. Clearly, however, a more concrete definition is required, especially for the 'industrial robot'. Such definitions vary as regards how general they are. The Japanese Industrial Robot Association (JIRA) divides the term 'robot' into six ever more sophisticated classes: manual handling devices, pick-and-place devices, programmable variable-sequence manipulators, robots taught manually, robots controlled by a programming language and robots which can react to their environments.

In Europe and the USA, the term 'industrial robot' does not include the first two classes of the Japanese definition. Instead it tends to be defined as, 'a reprogrammable device designed to both manipulate and transport parts, tools, or specialized manufacturing implements through variable programmed motions for the performance of specific manufacturing tasks' (which is the definition used by the British Robot Association – BRA). The definition used by the Robot Institute of America (RIA) is largely similar to that of the BRA, and talks of, 'a reprogrammable multifunctional manipulator designed to move material, parts, tools or other specialized devices through variable programmed motions for the performance of a variety of tasks'.

Figure 1.2 Example of hydraulically powered prosthetic system developed at University College, London

The term 'robot', as used in the West (and this book), does not therefore include such devices as remote-control manipulators (**tele-operators** or **telechirs**) or artificial **bionic** limbs or **prostheses** (figure 1.3) because although these devices may make substantial use of the same technology as robots they are essentially controlled by a human not a computer. It is particularly the inclusion of both pick-and-place devices and manual manipulators in the Japanese definition of 'robot' which makes the comparison of statistics on the production and employment of industrial robots in Japan, Western Europe and the USA so very difficult. However, in an effort to overcome this confusion the Japanese have coined the global term **mechatronics** to refer to the undergoing linkage of mechanics with electronics that is present in all these technologies.

The robot arm

Although mobile robots are one day likely to be common, as may even science-fiction type humanoid robots, for the present the state to which the industrial robot has evolved is best referred to as a 'robot arm'. Most are essentially a mechanical arm fixed to the floor, wall, ceiling or to another machine, fitted with a special **end-effector** which can be either a gripper or some sort of tool such as a welding gun or paintsprayer. The arm moves, by means of hydraulic, electric, or sometimes pneumatic **actuators**, in a sequence of preprogrammed motions under the direction of a controller which these days is nearly always microprocessor based and which senses the position of the arm by monitoring **feedback** devices on each joint.

Robots are usually taught a program either by an operator physically moving the arm through the desired sequence, or by guiding it through the motions by remote control. Some sophisticated robots can be programmed directly by telling the robot to move given distances in given directions. A few of the latest robots have sensory feedback and are, to a limited extent, able to react to what is going on in their immediate vicinity. Also, in an effort to increase the range of positions the arm can reach (its **working volume** or **working space**) a few robots have been placed on tracks or gantries, so providing limited mobility (as shown in a later chapter in figure 4.6). The range in size of robot arms is quite dramatic, with some miniature assembly robots able to move through only about ten cubic centimetres, while robots designed by Lamberton Robotics in Scotland can carry whole forgings weighing up to 1.5 tonnes through several cubic metres.

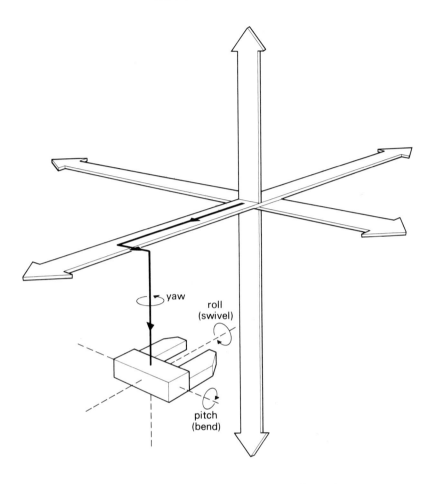

Figure 1.3 The six motions required to orient a gripper in any way at any point in space

Nevertheless, for the vast majority of industrial robots, it has been suggested that the equivalent human would be someone who was blind, deaf, dumb, had only one arm with a stub on the end, and had both legs tied together and set in concrete! Despite these incredible handicaps the robot arm has already made outstanding contributions to the manufacturing environment. Yet this is largely only because the environment in which it works has up to now been specially 'structured' for it, and is not identical to the environment which existed when a human did the 'same' job.

B

Different arm configurations

The essential role of a robot arm is to move a gripper or tool to given orientations at a given set of points. Mathematically, to be able to orient an object in any way at any point in space (Figure 1.3) requires an arm with six articulations (or **degrees of freedom** (DOF) – three **translational** (right/left, forward/back, up/down) to get to any point, and three rotational (**pitch**, **roll**, and **yaw**) to get any orientation. It should be noted that the addition to a robot of a gripper which can open and shut does *not* constitute an extra DOF for the robot, any more than the addition of a drill would.

All robots need the three translational degrees of freedom, but many dispense with one or more of the pitch, roll and yaw articulations (sometimes called 'bend', 'swivel' and 'yaw') and so save substantially on cost, often without noticeable loss of performance for such tasks as simple materials handling. The work envelope (all the points in space which can be touched by the end of the robot arm) varies in shape depending upon the actual configuration chosen for the design of the arm. One common structural classification of robot arms involves grouping according to the coordinate system of the three **major** (the translational) **axes** which provide the vertical lift stroke, the in/out reaching stroke, and the rotational or traversing motion about the vertical lift axis of the robot. Such a classification can distinguish between six basic types.

1   *Cylindrical coordinate robot*: In the robot shown in figure 1.4 the horizontal arm can move in and out parallel to the base, can move up and down the vertical column (remaining parallel to the base), and the whole base can swivel the arm and column around the vertical axis, so sweeping out a work envelope which is a partial cylinder. This corresponds to a mathematical coordinate system which specifies points in space in a similar fashion, and is therefore ideally suited for this type of robot. It should be noted that there is often so much computing power available in modern robots that the user can usually specify points in one of many different coordinate systems (such as the familiar x, y, z of the cartesian system) whatever the actual form of the robot, and the computer will do the conversion for him. Nevertheless, the **cylindrical coordinate** system remains the 'natural' system for this kind of robot, of which typical examples are manufactured by Fanuc, Prab and Seiko.

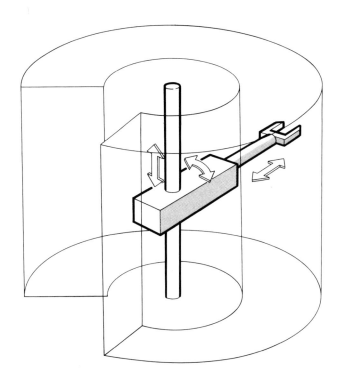

Figure 1.4 Cylindrical coordinate robot

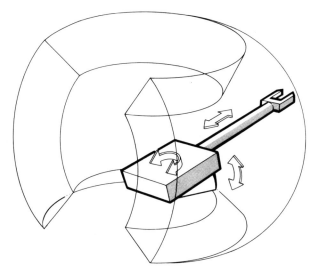

Figure 1.5 Spherical (polar) coordinate robot

2   *Spherical (polar) coordinate robot*: The robot in figure 1.5 has an arm which can move in and out and rotate on the base as before, but it utilises a pivoting vertical motion instead of a true vertical stroke, so sweeping out a partial sphere in space. This corresponds to the mathematical **spherical** (or **polar**) **coordinate** system. Typical examples of this kind of robot include the Unimate range manufactured by Unimation.

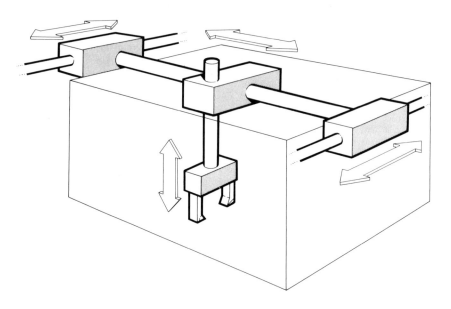

Figure 1.6 Cartesian (rectangular) coordinate robot

3   *Cartesian (rectangular) coordinate robot*: This type of robot, shown in figure 1.6, is configured with three mutually perpendicular traversing axes, consisting of an up/down column suspended from a beam on which it can move left/right, while the beam in turn is capable of forward/backward motion, so providing full x, y, z movement. This configuration is clearly ideally suited for direct usage of the mathematical **cartesian** (or **rectangular**) **coordinate** system. Examples of this kind of robot may have a gantry above, as in the IBM 7565 (originally RS1) assembly robot and the Olivetti Sigma, or may have the whole system 'on its side' like the DEA Pragma. Owing to the mechanical properties of such a configuration, it is a common choice where high precision is required, such as in certain assembly tasks.

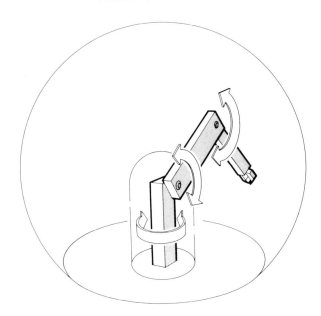

Figure 1.7 Revolute coordinate (jointed arm) robot

4 *Revolute coordinate (jointed arm) robot*: An example of this fourth class of robot, sometimes known as an **anthropomorphic robot**, is shown in figure 1.7. It consists of rotary joints called the 'shoulder' and the 'elbow' (corresponding to the human arm) all mounted on a 'waist' consisting of a rotating base which provides the third degree of freedom. This **revolute** (or **jointed arm**) configuration has the advantage of having a very large working envelope for its size, so minimising floor space requirements. Typical examples include those manufactured by Asea, Cincinnati Milacron and Unimation (the 'Puma' series).

5 *SCARA-type robot*: In April 1981 a fundamentally new structure of robot, developed at Yamanashi University in Japan, became commercially available for the first time. The so called SCARA (Selective Compliance Assembly Robot Arm), illustrated in figure 1.8, is similar to a revolute robot, but has the rotary joints in the horizontal rather than vertical plane, and uses a vertical lift axis attached to the end-effector. As such the design (based on a traditional Japanese folding screen called a 'byobu') exhibits properties of both revolute and cylindrical coordinate robots. Because of their stiff structure in the vertical direction, SCARA-type robots can bear far higher payloads than other assembly

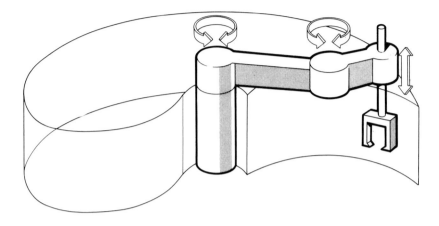

Figure 1.8 SCARA-type robot

robots – typically up to 30 kg. Various companies were involved in the initial SCARA research work, and several, such as Pentel and NEC, have brought out their own models. IBM has a licensing agreement to sell SCARA-type robots, such as the IBM 7535.

6 *Parallel robot*: A radical departure from conventional robot designs is the research robot GADFLY (GEC Advanced Device For assembLY), shown in figure 1.9. As can be seen from this figure it consists of a tool-mounting plate suspended from three pairs of rods; similar to an inverted aircraft simulator. By altering the lengths of the six rods it is possible to move the end-effector through all six degrees of freedom. What is particularly important about this design, however, is that because the articulations are not arranged sequentially as in other robots, the system can be very light, fast, and accurate. This advantage is at the expense, however, of a small working envelope. Such **parallel robots** may become increasingly common for light assembly work.

Figure 1.9 The 'GADFLY' – an example of a parallel robot

Robot classifications

In addition to classifying robots according to their arm configuration, various other terminologies are commonly used.

Fixed/variable sequence robots

Pick-and-place devices (also nicknamed **bang-bang machines**), although strictly speaking not robots at all, nevertheless are often referred to as **fixed sequence** (or **limited sequence**) **robots**. The stroke of each axis of motion is determined by adjusting mechanical end-stops, and sensors come typically in the form of limit switches which can only sense the end points, and none of the points in between. Such devices cannot be reprogrammed to execute a new task, but must instead be reset and adjusted as would a traditional automatic machine.

This is in fundamental contrast to true robots (that is **variable sequence robots**), which can immediately execute a different task or sequence merely by running a new program. To complicate matters however, there are a few pick-and-place type devices which can now switch in various different 'end-stops' under program control, so creating a 'grey area' in the definition. The ASEA 'MHU Senior', for instance, has eight mechanically set stops per axis, each of which is program-selectable, permitting complex sequences. In addition, of course, there is always an incentive for industry to refer to pick-and-place devices as 'robots', as this sounds far more glamorous and progressive.

Servo/non-servo robots

With variable-sequence robots it is necessary to be able to stop a particular joint of the arm at any point along its travel. There are basically two approaches to this problem. With the simplest method the controller merely sends power to the joint for as long as it estimates the arm will take to get to the desired position. Although with certain special electric motors (**stepper motors**) this approach can sometimes be satisfactory, on the whole such **open-loop** control with no feedback of information about the joint's actual position is grossly inaccurate – the arm might have got stuck and not have moved at all! Consequently, all but **educational robots** (or **teaching robots**), figure 1.10, make use of a second method, which involves placing a **servo-mechanism** on each joint which effectively checks both the position of the joint and the position in which the controller wants the joint, and then moves the arm until the

Figure 1.10 Example of a non-industrial 'educational robot'

two positions cooincide. Robots employing such **closed-loop** control are termed **servo-controlled robots** or simply **servo-robots**.

Point-to-point/continuous path robots

A common distinction is drawn between two different types of controller used in industrial robots. Many of the earlier robots only had sufficient computer memory to store discrete points in space which the arm had to move to, and during the movements between those points the path of the arm was not defined and often difficult to predict. Such **point-to-point (PTP) robots** are still very common and are perfectly adequate for such tasks as spot welding. As the cost of memory has come down, so the number of points which can be stored has increased, and many manufacurers use the term **multipoint control** if a very large number of discrete points can be stored.

For some tasks, such as paint spraying and arc welding, it is necessary for the path followed by the robot to be controlled at all times. Such **continuous path (CP) robots** in reality approximate a continuous line by splitting the path up into a very large number of separate points very close together. The positions of these points are either recorded during programming, or are calculated during the actual movement by filling-in (**interpolating**) between, for example two points to produce a straight line. These robots can be thought of as a natural extension of point-to-point systems, and there is in fact a 'grey area' in which multipoint control systems can approximate a continuous path system by not stopping at each discrete point but merely passing through them.

First/second/third generation robots

The **first-generation robots** are generally considered to be those 'deaf, dumb, and blind' robots which are currently most common on the factory floor. **Second-generation robots** have been around for some time in laboratories and are already in a few factories. Such robots, which may often *look* the same as their first-generation counterparts, can make use of varying degrees of sensory information about their immediate environments to modify their behaviour during a task (corresponding to the most sophisticated of the six classes of the JIRA definition of 'robot' mentioned earlier). Typical sensors include vision systems and **taction** systems (which provide a 'sense of touch').

Some people have called second-generation robots 'intelligent robots', but this term should really be reserved for the still more sophisticated **third-generation robots**, which are not yet even in the laboratories. In contrast, current research is really only concerned with 'common sense' robots! Nevertheless, research is indeed leading to so-called **intelligent robots** which will be equipped both with senses and the ability to recognise and have some understanding of objects in the outside world, and so to an extent, to have the capability of acting of their own accord.

As with all these definitions, there are 'grey areas' – *one* simple sensor does not make a second generation robot, it is necessary that the sensor(s) significantly affect the robot's operation, but how much is 'significant'? On top of this, even the accepted definitions vary: some authorities insist that the first generation was pick-and-place devices, so promoting all the other stages by one generation! As suggested in the final chapter of this book, it may well be that eventually only second-generation robots upwards will be considered 'true robots', with the first-generation being thought of a 'programmable devices'.

What can current robots do?

Current industrial robot arms tend to have been justified because of their untiring nature, predictability, precision, reliability and ability to work in relatively hostile environments. Robots frequently increase throughput and productivity, improve the overall product quality, allow replacement of human labour in monotonous and of course in hazardous tasks, may save on materials or energy, and yet throughout they have sufficient flexibility that they may be used for medium or even small batch production which traditionally has not been susceptible to automation. Such production is a major market: in a recent Pentagon study, for example, it was discovered that the vast majority of items purchased even by the military were made in lots of fewer than 100, and in the UK it has been estimated that roughly 75% of all metal parts made are produced in batches of less than 50!

Although robots do not as yet possess many of the important capabilities which come naturally to a human being, such as the ability to react intelligently to unforeseen problems and changing work environments, the ability to learn from experience, and the use of subtle hand/eye coordination, nevertheless, through the use of highly structured work environments (constantly becoming less restricted), robots are employed in a very wide spectrum of activities. Robots with grippers, or equivalent (such as the typical heavy duty robot shown in figure 1.11),

Figure 1.11 A heavy-duty industrial robot (a) as typically used (b) with covers off revealing hydraulic systems

are used in materials handling tasks such as **deburring, diecasting, fettling, forging**, heat treatment, **investment casting**, machine servicing such as loading and unloading, plastic moulding, and packing, palletising and stacking. Details of all these tasks are given in chapter 9.

Robot arms can be fitted with various kinds of tools in place of a gripper. These range from various types of applicators suitable for spray painting, adhesives, surface coating, powdering and sealing, to tools suitable for tasks such as drilling, countersinking, **nut running**, grinding, and sanding. In addition, the arm can be used for spot and arc welding, heat treatment and cutting using either flame or laser, and water jet cleaning. These applications are covered in chapter 10. It is interesting to note, however, that the original concept of a truly general purpose robot, capable of being used for almost any task from assembly to spot welding has now largely disappeared. Instead, robots are becoming slightly specialised, with 'paint spraying robots', 'welding robots', 'assembly robots', and so on, appearing as distinct types. Although usually capable of performing many other tasks, each robot design is often *particularly* suited for a specific niche in the market.

Finally, it is salutary to remember, given the large number of potential 'steel collar' replacements for human workers, that a robot can only ever, by definition, replace someone who is 'working like a robot'. In not such a long time it may well be that robots will be able to replace humans not merely in boring, repetitive, or dangerous work, but in activities which were previously considered to require substantial skill, learned over a lifetime. This likelihood understandably causes great concern to many, and robots have inevitably been equated with unemployment. This very important topic is dealt with in detail in chapter 13. Nevertheless, although with such advanced robots the (admittedly convenient) argument that robots were merely removing humans from unpleasant jobs could no longer apply, many people argue that under those circumstances it would surely seem degrading for mankind to continue to spend his life simply acting out the role of a machine.

2

From Ancient Greece to factory grease
THE EVOLUTION OF ROBOTICS

Development of modern robotics

Although it is only very recently that people have been able to specialise exclusively in the field of robotics, nevertheless research which we would now classify as robotics has been conducted for several decades under the different guises of such fields as **artificial intelligence** (AI), Computing and Control, and Mechanical and Electrical Engineering. It is always tempting to look at the **state-of-the-art** in a given subject and tacitly accept a form of 'technological determinism' which assumes that the field could only ever have developed in that way – possibly faster or slower, but at least always passing through the same stages.

Yet robotics is a highly multidisciplinary subject drawing on computing, mechanics, electrical engineering, mathematics, fluid engineering, metallurgy, control engineering, psychology, industrial sociology and so on. There are so many (sometimes conflicting) forces acting upon the developing field that the present form of robotics, with its almost total preoccupation with industrial robotics, is far from the only one that could have occurred. In this chapter we will look not only at what robotics-related work has led to the current state-of-the-art within the field, but also at the concurrent development of automation and the resultant emergence of industrial robotics, finishing with an overview of the present state of robotics worldwide.

While the automata mentioned in chapter 1 anticipated modern robots by moving like humans or animals, so there developed many systems which, while not looking like 'robots', nevertheless started to act like them. The first mechanical calculating machines were built in the seventeenth century by Pascal and later by Leibniz, and the concept of programmable machinery followed soon after in the 1720s in the form of developments in France which eventually led to the constuction of looms in which the lifting of the correct threads to produce the required complex patterns in the cloth was controlled by a band of punched cards, perfected by Jacquard in 1801.

From 1823 Charles Babbage was able to use his mechanical 'difference engine' for the automatic computation of tables for navigation, insurance and astronomy. This first design was only capable of producing such tables rather than acting as a general–purpose calculator. However, Babbage's designs for a more sophisticated mechanical computer which he called his 'analytical engine', would have allowed true programming of a required task by means of punched cards similar to those used in the Jacquard loom. Unfortunately, although these revolutionary designs have subsequently been verified, during his lifetime the attainable mechanical tolerances were not good enough to actually construct working machines. It is interesting to note, however, that *electrical* computers based on relays *could* probably have been constructed in the late nineteenth century – only nobody got round to doing it! Meanwhile, in the USA, in 1873, C M Spencer began the programmable machine tool industry by building his 'Automat' lathe for producing screws, nuts and gears, which was automatically controlled by guides fitted onto two rotating drums he termed 'brain wheels'. One drum effected the movements of the workpiece, while the other controlled the sequence of cutting operations. About the same time, back in England, James Clerk Maxwell presented a scientific paper *On governors* which described the operation of the flyball governor used for regulating the speed of steam engines. In his detailed analysis, he was among the first to conduct a systematic study of the concept of feedback, which is now fundamental to the control of robots.

In 1890, Dr Herman Hollerinth, in his role as statistician to the US Bureau of Census, employed punched cards to record the census data for each person, allowing automation of the processing of data. Forty years later, in 1930, the first **analog computer** was constructed, and the urgency of the Second World War precipitated the construction of a mathematical basis of feedback control for use in high-speed aiming of artillery using radar tracking. In 1944 the first fully automatic **digital computer** was built by IBM. With 750,000 components and 500 miles of wiring, this leviathan could add two numbers in a third of a second – but division took 10 seconds.

Also during the war years, various programmable paint spraying machines were patented. The time was ripe for the union of the physical manipulative ability of mechanical systems with the controlling sophistication and flexibility of the fully electronic computer , which for years was thought to have been first constructed at the University of Pennsylvania in 1946 in the form of the 30 ton ENIAC, which contained 18,000 valves and took up 1700 square yards of floor space. In fact the British equivalent, aptly named 'Colossus', and built at Bletchley Park, Buckinghamshire, was operational in 1944, and was successfully applied to break the Germans' 'Enigma' code generator. However, the secrecy surrounding the project prevented its full details becoming public knowledge until the late 1970s.

In 1948 Norbert Weiner published *Cybernetics, or control and communication in the animal and the machine* and so popularized the whole field of control (feedback) and communication theory, which has since consequently been termed **cybernetics** – a name actually used by Ampère over a century before and derived from the same word as 'governor', the study of which had set the whole field going. At the same time W Grey Walter was demonstrating his 'Machina Speculatrix', little mobile 'tortoises' which used mechanical contact to find their way round objects, headed towards light and even returned to recharging 'hutches' when they grew 'hungry for electricity'.

Also in 1948, the British built at Manchester University the first computer which actually stored its program of instructions. Nevertheless, most significant of all in this auspicious year was the discovery of the transistor. Its small size, low power requirements and high reliability made it destined to have a revolutionary effect on the computer. Up till then, even with unlimited funds and with no restriction on size, it would have been impossible to build a powerful computer, simply because valves were so inherently unreliable that any computer with more than a certain maximum number in it would never have all the valves working at the same time! In 1950, the UK company Ferranti produced their Mark 1 Star, which was probably the world's first commercially available computer. Although far smaller than the early ENIAC, it nevertheless produced heat equivalent to 27 one-bar electric fires! From the early 1950s onwards the electronic computer developed swiftly with substantial improvements in memory and overall system design, while in 1951 a general purpose digital program storage device was patented for controlling automatic machine tools.

In 1954, George Devol applied to patent a design for what is generally considered to be the first 'industrial' robot. It consisted of a general purpose manipulator with a playback memory and point-to-point control. Two years later, at a now famous cocktail party in the USA, Devol met up with 31-year old Joseph Engelberger who was working as an aerospace engineer. As the party progressed, and talk became wilder, the two men came up with the idea that they should actually set out to build flexible machines for factory automation. In 1958 Devol's robot was actually built. Later, Devol sold his early patents to Condec, out of which grew the company Unimation Inc. (from 'universal automation'), headed by Engelberger, which over a decade later in 1972 was the first firm anywhere to devote itself solely to the manufacture of robots. It did not show a profit until 1975 . . .

In Summer 1956, the term **artificial intelligence** (AI) was popularised by John McCarthy (who has the distinction of having founded two of the world's leading AI laboratories), at a conference held at Dartmouth College, New Hampshire. On the industrial front, in 1957, a pick-and-place device was developed by the Planet Corporation, USA, which could be mechanically 'programmed' by means of cam-wheels. Even at this stage, however, there were less than 2000 computers in the USA and only 140 in Western Europe! By the early 1960s the major features of the first-generation robots had been patented and the first industrial installation was carried out by Engelberger and Devol in 1961, a year before Unimation was officially launched!

This first robot was used to unload a die-casting machine at a General Motors car factory in Trenton, New Jersey. The robot actually looked quite like its modern descendant shown previously in figure 1.11. It is interesting to note that the mechanical portions of the first 'Unimate' had been available some years previously – it was the controller which held things up. This is understandable when it is remembered that computers were still very large and far too espensive for dedication to a robot – there were still only about 7000 computers in the USA and 1,500 in the whole of Western Europe.

However, by the mid 1960s robots had the ability to branch to one of several recorded programs dependent on an external stimulus, and continuous-path control had been developed. At about the same time robotic research labs sprang up at MIT, the Stanford Research Institute, and the University of Edinburgh, and in the late 1960s and early 1970s, SRI built various versions of 'Shakey' – an 'intelligent' mobile robot equipped with television camera, rangefinder and bump detectors, which could perform such tasks as finding its way round obstacles, and pushing designated boxes together in specified patterns.

Early robots, although incorporating computer hardware such as memory, nevertheless consisted of electronic logic components which were **hardwired** to perform a specific set of tasks, with the electronics effectively duplicating the functions of the earlier **hard automation**. In 1968, however, at MIT a robot was connected to a general purpose PDP-6 computer, and so the first truly flexible robot was born. Later, in 1974, the first commercially available minicomputer-controlled 'T3' robot was produced by Cincinnati Milacron. Such **softwired** robots are far more powerful than machines with specialised electronic circuits – they can work in several different coordinate systems, react to sensors and make use of sophisticated teaching techniques.

Robotics research, although not always fully supported, continued to produce valuable direct benefits throughout the 1970s. Yet it suffered a major setback, at least in the UK, when in 1973 Sir James Lighthill produced a report for the then Science Research Council (which funds most University research in the UK) on the prospects of research into artificial intelligence. He concluded (it now appears, utterly wrongly) that AI research was doomed to failure. This unfortunate finding effectively cut off nearly all funding overnight. Although AI researchers were not the only people involved in robotics, the Lighthill report undoubtedly severely damaged any chances that the UK might have had of being a world-class contributor to robotics research during the 1970s.

In the USA, however, an academic, Vic Scheinman, was building small robot arms, first at Stanford and then at MIT, and, after forming the company 'Vicarm' to promote his designs, sold it to Unimation who proceeded to develop the PUMA arm (one of the first commercial assembly robots – figure 2.1) based on his work. At Stanford the first computer-integrated robot assembly station was developed in 1973, to put together a ten component water pump. In 1976 the Draper Labs developed a **remote centre compliance** (RCC) device which acted like a floating wrist to accommodate slight inaccuracies when positioning parts during assembly; this was in the same year that the Viking 1 'robot' capsule, built by NASA, landed on Mars.

Apart from occasional developments in the mechanics of robots (such as the RCC), the majority of progress in the field was by now advancing hand-in-glove with the staggering improvements in computing. Since the mid 1960s, the memory density of computers has about doubled regularly every year – that is 1,000 times the power in ten years, 1,000,000 times in twenty! The development in the late 1960s, of **large-scale integration** (LSI), placing thousands of small components on one speck of silicon, made available previously unthinkable computing power for a reasonable price, and so started the 'third generation' of computers since the early 'valve days'.

Nevertheless, it is worth remembering that the mere ability to miniaturise electronics does not automatically make them cheap. Producing the first of a given design of integrated circuit is unbelievably expensive. It is only because the function of these 'chips' is *so* flexible and their potential applicatons are *so* widespread that it is possible to sell millions of the *same* chip (so spreading the heavy design costs over a very large production volume). There is a close analogy here with the advantages of the inherent flexibility of robots themselves.

Computer languages could now become internally far more complicated, so that externally they appeared far more 'natural' to the user.

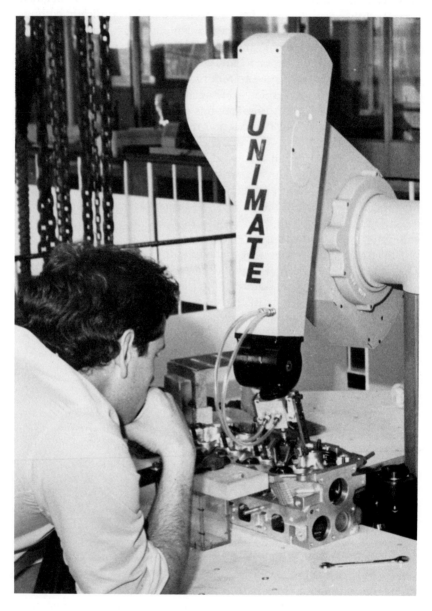

Figure 2.1 The Unimation 'Puma' – one of the first commercial assembly robots

Sophisticated **pattern-recognition** systems allowed limited vision, at least in the laboratory. By the beginning of the 1980s work had begun in earnest on the second generation of robots which had sufficient computing power to be able to modify their behaviour in response to sensing the immediate environment, and so were potentially capable of accomplishing such complex tasks as arc welding and assembly operations. Such work involves using computers based on **very large-scale integration** (vlsi) with, for example, each memory chip able to hold tens or hundreds of thousands of computer-bits of information (see figure 2.2).

(a)   (c)

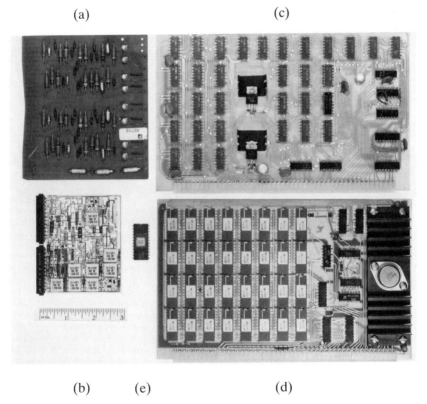

(b)   (e)   (d)

Figure 2.2 Examples of different stages in computer-circuitry evolution (a) Board from an 'Atlas' computer c.1950. Descrete transistors. Stores 4 bits. (b) Board from an IBM computer – late 1960s. Early integrated circuits each about equivalent to board 'A'. (c) Board from an early microcomputer c.1975. Stores ~32 000 bits. (d) Later board from the same microcomputer as 'C' c.1978. Stores ~256 000 bits. (e) A single chip can now hold more than the whole of board 'D'.

The complexity of the interconnections on such chips is staggering: greater than a street map of an imaginary city stretching over the whole of the USA!

In addition to such tremendous advances in economic computing power (which also fired a whole new 'technological revolution'), because of a substantial increase of industrial interest in robots, in part brought about by the need to increase efficiency due to a world recession, several programmes were started to discover how robots should best be installed, when they were cost effective and how to design and control large robotic installations. Details of such ongoing research are covered in the last two chapters of this book.

Development of industrial robotics

As pointed out above, it was never inevitable that the industrial applications of robotics should have been its first major market. It is all too easy to think of robotics as being a currently very fashionable subset of industrial automation which, after the initial excitement has died down, will become just yet another tool at the production engineer's disposal. Yet this is only true of that part of robotics which is 'industrial' robotics; there are potentially very many 'nonindustrial' applications for robotics which may even one day overtake industrial robotics in importance. If, 30 years ago, by some quirk of history, mankind had been very adept at building computers yet comparatively poor at constructing mechanical devices, we might now have large numbers of 'intelligent' **domestic robots** and be wondering how long it would take before this robotics technology could be adapted for practical industrial use, in much the same way as the Ancient Greeks used the principles of the steam engine to operate gimmicks in their temples, yet it took the best part of two millennia before the same principles were put to practical use.

Modern industry has grown out of the mechanisation of the industrial revolution which was fired by the development by James Watt of his steam engine in the latter half of the eighteenth century. With mechanised production came the ability to make parts which were almost identical – so much so that in the early nineteenth century Eli Whitney was able to produce 12,000 muskets for the US Government in which all the same parts were interchangeable between guns. Mass production had truly arrived, and by the late nineteenth century the constant demand for improved production rates had led to several refinements to metal-cutting machines which culminated in automatically controlled machines, such as Spencer's automatic lathe mentioned above.

The seemingly insatiable desire for the newly invented motor car created a demand for better machine tools to manufacture them with, and the development by Henry Ford of the **production** (or **transfer**) **line** meant that by 1914 his company was producing over 1,000,000 model T cars a year. However, although such automation (which later in some cases became extremely sophisticated and was called **Detroit automation**) is suitable for very large volume production, with reduced labour costs, and low unit cost, even so, it requires long **lead times** (a major change in design may require several years to set up the line) and so is inherently inflexible to evolving technology and requirements.

For a long time medium- and low-volume **batch production** had to be undertaken using conventional manually operated machine tools, but in the late 1940s John T Parsons suggested a method for automatically guiding a **milling machine** by means of coded punched cards in order to machine the complex shapes of the helicopter blades he was working on. Then in 1949 the US Air Force commissioned the MIT Servomechanisms Laboratory to develop such a **numerical control** (NC) machine. A decade later, NC systems were being successfully used in production, and work had begun on developing a programming language for their use called APT. Nevertheless, as touched on in chapter 1, NC or even CNC systems are not thought of as 'robotic', because they do not exhibit the flexibility of *type of operation* that true robots do.

Although the application of numerical control to specific machines grew in range from automatic fabric knitting to blueprint drafting, the newly available industrial robot of the 1960s, which was in effect a general purpose handling machine, had an uphill struggle to find cost-effective applications. Whereas NC machines tended to replace craftsmen operating manual versions of the same machine, robots were capable only of 'unskilled' labour which was still relatively cheap. In addition, it was not at all clear for which jobs a robot was best suited, as there was no single type of machine for which it was an automatic replacement.

As the sophistication of the robot controllers improved however, and techniques were developed for structuring robot surroundings in ways which to an extent compensated for robot inadequacies, so new applications such as paint spraying became cost effective. In the 1970s, industrial robotics came to be considered as one of the aspects of an overall philosophy of **computer aided manufacture** (CAM), and with ever increasing labour costs and pressure to remove workers from hazardous or unpleasant jobs, together with an insistence from the consumer market for wider diversity of product styles, interest in robotics grew in industry. As the world recession of the late 1970s struck, so academic interest rapidly turned into a desperate fight to increase productivity while trimming manning levels.

Mechanical engineering requirements for industrial robots had never really been particularly demanding, and could more than be catered for by existing technology. The restraining technical factor however was primarily in control, yet even so, the available computing power for cost-effective industrial robots was just about sufficient to allow them to start to integrate with the factory environment, and take over many materials-handling tasks and a few 'semiskilled' jobs such as painting and spot welding. It has been pointed out that there is an analogy here to historical changes in the approach to warfare. Since the eighteenth century, far greater autonomy has been given to the local commander and even the individual soldier, by training a high level of intelligence into each man (rather than simply dumb obedience). As with robots on the factory floor, such autonomy requires sophisticated intelligence gathering and communication systems, which simply were not available before.

Towards the end of the 1970s, the promise of increased computing power in only a few years, prompted many countries to invest heavily in second-generation robotics research in the hope that the implementation of such technology would help to 'pull' them out of recession. In addition, plans by the world computing community to develop **fifth-generation computers** within a decade, stimulated research into artificial intelligence and related topics. So, due to a combination of historical factors, in the early 1980s there was a tremendous boost given to robotics work in general, and to industrial robotics in particular.

Nevertheless, this must of course also be seen in the more general context of the overall technological revolution started in the 1970s, largely based on the integrated circuit. This was producing consumer products such as miniature calculators, personal computers, video games and digital watches. Thus every 'man in the street' was directly affected by the new high technology, and there was consequently a high level of awareness even among the more conservative professions that 'something important was happening'. **Information technology** (IT) became a catchphrase as computing techniques infiltrated even offices of traditionally 'low-tech' firms in the form of word processors and electronic mail. So, to an extent, robotics was able to take advantage of this new awareness of high technology, and enter markets which might otherwise have required substantially more persuasion about potential benefits.

Robotics worldwide

Despite the origination of robotics in the West, the country which undoubtedly took most advantage of first-generation robots was Japan. One of the factors involved in this may have been that, in contrast to Western practice, many members of Japanese firms tend to have a 'job for life', so that management may have felt able to make longer term, longer payback, higher risk plans involving the use of robots than their Western counterparts would have felt were 'safe'. Similarly, the introduction of robotic technology held no fear for the Japanese workforce because they were still guaranteed a job somewhere in the factory.

Some Japanese claim that there is a historically based 'team spirit' prevalent in Japan, because before the Meiji period of the 1890s *everyone* had to work together harmoniously in irrigating the paddy fields to avoid disaster and starvation. Whatever the validity of this belief however, there is the undeniable success of far-sighted consistent Government support for robotics, and of the Japanese ability to take maximum advantage of other countries' inventions. Interestingly though, the first Japanese company to become heavily involved in robotics was Kawasaki – yet that was as late as 1968, a decade behind the US lead.

Although, owing to the different definitions employed by various countries, it is very difficult to compile accurate worldwide surveys of 'robot populations' in different areas, many attempts have been made, from which an overall picture emerges. Even by 1970 there were only a few hundred robots throughout the whole world, with 200 in the USA and 150 in Japan. By the mid 1970s, however, the world population was closer to 4000, and by 1984 was about 37,000. At the present time, Japan appears to have more installed robots than any other country,

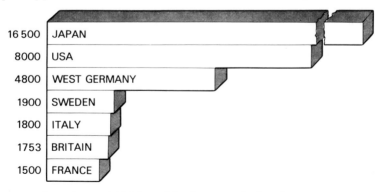

Figure 2.3 Estimated 1984 world robot population figures

even when the differences in definitions are taken into account. These differences, however, can make a substantial difference: in 1981, for instance, Japan was claiming to have 70,000 robots, whereas by the West's difinition of 'robot' the Japanese only had about 8,000 – interestingly suggesting that a larger proportion of the Japanese devices were of the low-technology type than was the case in the West.

Japan is followed in robot population by the USA, then West Germany, followed by a group consisting of Sweden, Italy, the UK, and France, all with about the same robot populations. The order has remained about constant for several years, although all countries have been installing at an accelerating pace. In 1984, the British Robot Association estimatd the world robot population figures to be as in figure 2.3. This was interestingly the first time that, at 12,500, the number of Western European installations (excluding Eastern European countries and the USSR) substantially overtook the number in the USA. Reliable estimates from the USSR and allied countries are very hard to obtain, although some sources place them at 3–6,000, suggesting that the European Continent as a whole contains more robots even than Japan. It is interesting to note that the highest *density* of robots per number of workers is in fact in Sweden, rather than Japan.

Although Japan undeniably leads the world in actual numbers of installed robots, in robotics research they seem to have little or no superiority. Among the world leaders in research should be included the USA, UK and West Germany as well as Japan. Similarly, there are many large robot manufacturers outside Japan, some of which have specialised in the field for several years. Recently there has been a proliferation of new robot manufacturers (with over 100 now in Japan alone) and it is debatable how long the market will sustain such a wide diversity of suppliers. In many countries, just a few firms may supply over 80% of the robotics market, leaving only a 20% share for all the other suppliers to fight over.

Many predict a substantial 'shake-out' in the industry, leaving a comparatively small number of large manufacturers (with the required combination of product technology, systems and applications expertise and marketing skills) to serve the whole world market. A report by Arthur D Little, for example, recently predicted that the worldwide robotics market would grow at an average annual rate of 28%, rising from $175 million in 1983 to $600 million in 1987 and (in 'today's dollars') $2 billion in 1992 – equivalent to maybe 50,000 robots installed in that one-year period!

A report by Frost and Sulivan estimated that in 1982, Japan produced 7000 programmable robots, the US produced 1800 and Europe 3200.

Among the more notable robot suppliers are: Unimation, now part of Westinghouse, wholly involved with robotics and recognised as the original pioneers of the industry; ASEA, a very large Swedish electrical engineering company with long involvement with robotics, and now Europe's leading robot manufacturer; Cincinnati Milacron, USA, the largest machine-tool manufacturer in the Western-world, which built some of the early computer-controlled industrial robots; IBM, the giant computing multinational which has recently penetrated the robot market and is likely to have a substantial influence upon it; as well as many Japanese firms such as Hitachi (which has formed many international licence agreements to sell its products) and Fujitsu Fanuc (which also makes controls for machine tools and is building up great experience in **unmanned factory** design).

In addition, there is a growing tendency for large companies to attempt to build up a 'robotic presence' by taking over some existing robot producers and forming licence agreements with others. Such licensing agreements are only likely to be a short term measure. Indeed it is already being predicted that in only a few years many large manufacturers who once joined with technically skilled producers, will once again break away from such overseas partners.

Of all the current worldwide industrial robot applications, until now almost all have employed first-generation technology, and the most common uses have been for surface coating, spot welding, parts handling and machine servicing. It is interesting, however, that different countries appear to use substantially different proportions of their robots for the same tasks, and, as noted above, the majority of the Japanese 'robots' seem to rely on low technology. Recently, however, it is noticeable that worldwide there has been a dramatic increase in the application of robots for arc welding and assembly, tasks which largely require second-generation robots. Indeed, it is predicted that towards the latter half of the 1980s, assembly may take over as the *major* robot application.

Spot welding, on the other hand, looks like soon largely saturating its predominantly automotive market, so that the number of new such installations is likely to drop substantially towards the late 1980s. Even relatively cheap and simple teaching robots are finding increased application outside the educational environment, as some of the more sophisticated versions become sufficiently advanced, some believe, for limited application in commercial use.

It is ironic that a major use for robots continues to be on the car production line, where the inherent flexibility of a robot can never really be fully utilised. Indeed, Creative Strategies International, a market research firm, estimated that 75% of the 1980 US 'heavy duty' robots were in the car industry! It is a further sad irony that in the poorer countries of the Third World, where most manufacturing is of the medium to small volume ideally suited for robotisation, they are at the moment utterly unable to provide economic justification for using any robots whatsoever.

Part II
ROBOTICS TECHNOLOGY

One of the major attractions of working in robotics is its multidisciplinary nature. Yet, of course, this is also its curse: however much of an expert one may be in any specific branch of technology, suddenly one finds one is a mere beginner in several of the other branches which are equally important to robotics! In the six chapters in this section, the various types of technology employed in both robots and robotic equipment are explained to a depth at which someone 'outside' a given technical field can at least begin to appreciate some of the difficulties faced by his counterpart within the field.

3

Strong arm tactics

ROBOT MECHANICS – I
STRUCTURES AND DESCRIPTION
TECHNIQUES

Manipulator anatomies

It was stated in chapter 1 that of the basic types of structures for industrial robotic manipulator arms, there are four which have been around for some time: CYLINDRICAL (rotation–translation-translation), SPHERICAL (rotation–rotation–translation), CARTESIAN (translation–translation–translation) and REVOLUTE (rotation–rotation–rotation). Nevertheless, it is important to stress that these are not the only forms which are potentially useful. For instance, it was pointed out that recently some manufacturers have started producing SCARA-type arms which use a rotation–rotation–translation structure for the three major joints in which, unlike spherical systems, both rotations are in the same plane.

Of course, it is also possible to have more than one actuator working in *parallel*, as in the Gadfly robot shown earlier in figure 1.10. Nevertheless, in practice, only configurations of actuators linked *serially* tend to be used for industrial robots. Even so, if all the various possible manipulator structures (for the three major axes) are considered, then once those which would not actually 'sweep out' a useful working envelope have been disregarded, there still remain 37 different possible structures for systems! For example, it is easy to imagine a new design of rotation–translation–rotation robot such as that shown in figure 3.1, where a forearm on the end of a horizontal telescopic support can rotate in the vertical plane, while sweeping out an envelope by rotating the whole system around a vertical axis.

It is not really surprising that, up till now, so few of the potential configurations have been generally used. When available computing power is low (as it was until recently), it is convenient to employ a physical robot configuration which directly corresponds to one of the

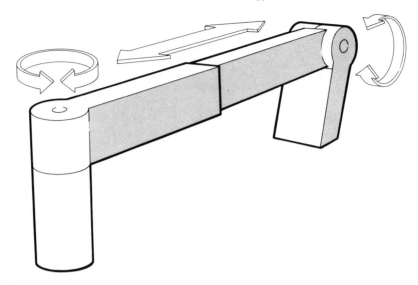

Figure 3.1 A hypothetical form of rotation-translation-rotation robot

common mathematical coordinate systems (such as spherical or cartesian). If this is done, it is then straightforward to specify coordinates in a way which both the user and the robot can understand. Otherwise, sophisticated computation must be employed to convert coordinates from one type to another. However, such conversions may be required very quickly, several times a second, and so unless significant computing power is available, the conversions cannot be performed in **real time**.

In addition to the three degrees of freedom (or **axes of motion**) which all robots must have, the 'wrist assembly' may consist of one to three additional degrees of freedom, depending on the applications for which the robot is destined. Such wrists are often designed in a modular fashion so that the same model of robot can be fitted with various numbers of degrees of freedom. It has become quite common, for instance, for owners to send their five axis robots back to the factory after a few years, so that they can be upgraded with a sixth axis, so opening up the range of suitable applications for the robot to include such areas as assembly.

Wrists themselves, as far as possible, tend to be arranged so that each axis of rotation passes through the same point. A common method of arranging the yaw and roll movements is to employ a **differential** as shown in figure 3.2. Shafts A and B can be driven independently of each other, and an end-effector is fastened to the end of shaft C. If A and B

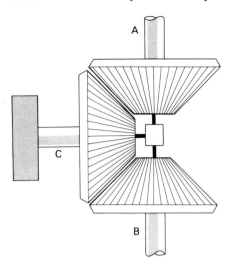

Figure 3.2 A differential gear mechanism

move at the same rate but in opposite directions then C will rotate about its major axis, resulting in roll of the end-effector. If, however, A and B both move in the same direction (and at the same rate as each other) then C will not rotate about its major axis, but will instead pivot about the axis of shafts A and B, causing yaw of the end-effector. Unequal speeds of A and B will result in simultaneous yaw and roll movements.

In practice, for a given application, some of the common structures of robot tend to be more often used than others. Of all types, revolute

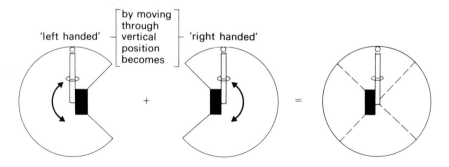

Figure 3.3 Plan of a jointed-arm robot, showing how the 'left'- and 'right-handed' configurations sweep out work-envelopes which overlap

arms are probably the most popular, especially since cheap computing has allowed them to be programmed initially in any coodinate system. Their large **working volumes** for their size has made them ideal for tasks such as paint spraying and arc welding. Many such jointed-arm robots have work envelopes which consist of a complete spheroid round the centre column, because, although the base joint can usually only rotate less than a complete revolution, the arm itself can 'flip' right over through the vertical position, and so access that sector of the sphere which would otherwise be unreachable (as in figure 3.3).

Although revolute arms are very suitable for assembly tasks, SCARA-type designs are increasingly being developed for such work because of their ability to appear stiff in the vertical direction while 'giving' slightly in the horizontal plane. Such **structural compliance** is dealt with further in chapter 8. Cartesian robots can also be used for assembly purposes, because, particularly when the system is suspended from a gantry above, the work envelope can be very large.

Heavy loads can often be carried by cylindrical robots, because it is possible to make the central column very rigid, by holding it both from

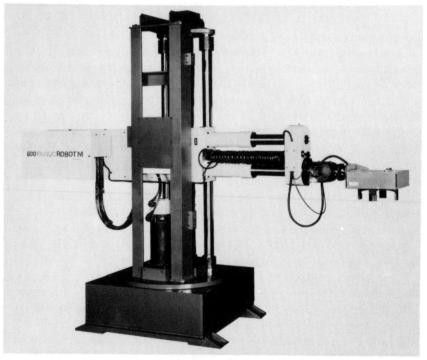

Figure 3.4 Example of a rigidly designed cylindrical robot

above and below, as shown in the Fanuc robot in figures 3.4. Similarly, because of the in-and-out horizontal movement needed for many machine-tending operations, cylindrical configurations are often ideally suited. Finally, polar robots can offer great flexibility to access locations, yet can frequently transport far heavier loads than the equivalent revolute device. This has resulted in polar robots, such as the ubiquitous Unimates, being heavily utilised for such tasks as spot welding and materials transfer.

Transformations

Whatever the configuration of a particular robot, if any computation is to be performed involving its given structure, then mathematical techniques must be employed which permit the description of the varying locations of the different robot-arm joints. Such techniques allow the absolute position and orientation of the robot end-effector to be determined from the relative positions of the robot joints, and (still more difficult) the necessary joint locations needed to place the end-effector in a required position. These sorts of computation are necessary for sophisticated control of the robot arm.

Each rigid section of a robot manipulator arm can be thought of as having its own private coordinate system locked around it. In this way, it is only necessary to specify the location and orientation of that coordinate system relative to some **base system** to in fact specify the absolute position of that section of the robot arm. Assuming that the coodinate system used is cartesian (although the approach can be modified to accept other systems), then figure 3.5 might represent the absolute position of the robot-section under consideration (for instance the end-effector) with regard to the chosen base system. The base system might, for example, be chosen so that the origin was at the centre of the base of the robot, with the X-Y plane horizontal, and the Y axis pointing away from the front of the robot base. The equivalent coordinate system around the required section of the robot arm can be thought of as 'floating' within the base system.

To actually describe the position of this second system in figure 3.5, much can be achieved merely by specifying the direction of the **vector** from the origin O_B of the base system to the origin O_R of the other system; however, this will not provide any information about the orientation of the second system. Mathematically, orientation can in fact be specified by using a **matrix** consisting of three rows of three columns of values. The three values in a given column in effect repre-

C

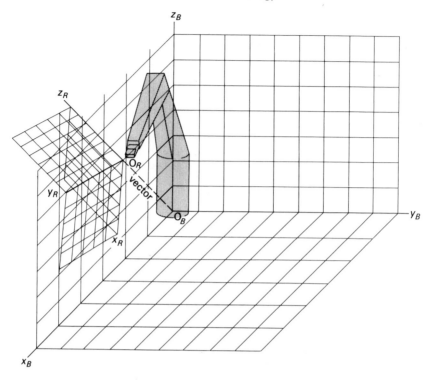

Figure 3.5 The coordinate system of a robot end-effector, shown within the coordinate system of the robot base

sent the *X Y* and *Z* orientations of a particular axis of the second system, as if it were a straight line in the base system. By specifying in this way the orientation for each of the three axes of the second system (taking three columns of the matrix) the orientation of the whole of the second system can be uniquely specified.

By combining the information in both the position vector and the orientation matrix it becomes possible, in principle, to start with the coordinates of a point in the system 'locked round' the section of the robot under consideration, and then **transform** them into the equivalent coordinates of the base-system. In practice this is performed by using a **homogeneous transformation** which consists of a matrix of four rows of four columns, easily constructed (as shown in figure 3.6) by writing the orientation matrix next to the position vector, and then writing 0 0 0 1 underneath them! In this form the transformation can be accomplished merely by multiplying the coordinates of a point in the second-system

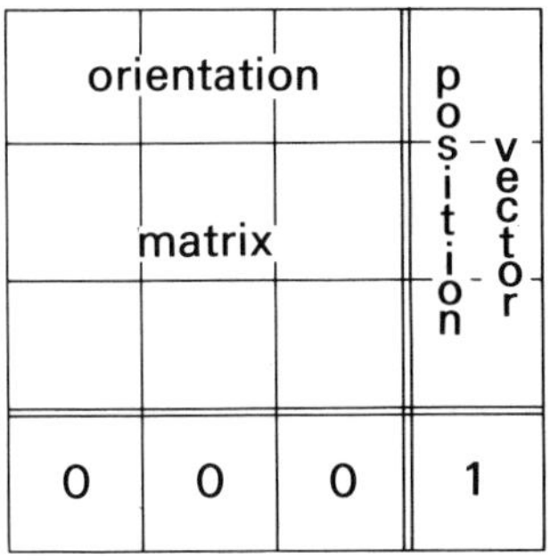

orientation			p o s i t i o n	v e c t o r
	matrix			
0	0	0	1	

Figure 3.6 The construction of a homogeneous transformation

(in the form $X, Y, Z, 1$; where the '1' is a scaling factor) by the homogeneous transformation matrix using **matrix multiplication**.

Clearly, in practice it would not always be convenient to specify a set of secondary coordinate systems all in terms of the base system. For a start, robots do not consist of a series of independent sections 'floating' in the base system – the sections are all linked one to another in a serial fashion. Secondly, descriptions of positions of objects (such as parts on a **pallet**) which are not connected to the robot itself, may remain stationary relative to each other, but not realtive to the base system.

To take account of such factors, homogeneous transformations can be employed to describe the positions and orientations of two coordinate systems relative to each other, both of which are different from the base system itself. In this way, the parts on the pallet, for example, could each be described, once and for all, relative to the pallet. As the pallet was subsequently moved, only the transformation relating the pallet to the base system would need to be changed. By employing the **relative transformations** which described the relation of the objects to the pallet, the absolute coordinates of the objects could be determined comparatively easily wherever the pallet was moved.

By stringing a series of relative transformation calculations together in this way, it is possible to work along a 'chain' of different systems in which only the relative positions and orientations of neighbouring systems are known. Of course, the most important chain is usually that of the robot arm itself. By using relative transformations, it is possible to work outward from the base of the robot (fixed relative to the base system) through each joint in turn, until the absolute position (with respect to the base system) of the end-effector itself is determined. This approach is dealt with further after the next section.

Euler angles

It was stated in chapter 1 that, in addition to the three translational degrees of freedom, a robot needed three more rotational articulations (pitch, roll and yaw) in order to orientate its end-effector at any angle. We will now consider this in more detail. In an unpowered glider the pilot basically has three controls: movement of the joystick forwards and back to alter the angle of the elevator surfaces on the tail, movement of the joystick from side to side to move the aileron surfaces on the trailing edges of the wings (it is only for the sake of convenience that these last two controls are both operated using the same stick) and the foot-pedals controlling the rudder.

Pushing the joystick forward puts the glider into a nosedive; pulling the stick back causes the nose of the glider to pull up. This is, in effect, the pitch control of the glider, causing it to 'pitch' in the same direction of movement as a boat pitching in a rough sea (figure 3.7a). Moving the control stick from side to side causes the whole glider to rotate around the axis of the fuselage. This is the Roll control. Finally, pushing on the rudder pedals causes the nose of the glider to slew to the side, the whole glider pivoting in a horizontal plane to point on a new course. This movement is known as yaw (figure 3.7c).

In practice, there are secondary effects to these controls which make it more efficient, for instance, to bank the glider into a turn rather than merely use the rudder. Nevertheless, in principle, the pilot has controls which allow him to rotate the glider independently in any one of three mutually perpendicular directions. Using the same convention as the pilot for roll, pitch and yaw, it is possible for the roboticist to specify the actual angles moved through for each of the three types of rotation, and so to specify uniquely the orientation of one coordinate system relative to another. When relative orientations are specified in this way, the three rotations are commonly referred to as **Euler** (pronounced 'oiler') **angles**.

Of course, in specifying the three Euler angles, the sequence of rotations is most important. Imagining a glider again, a 90° clockwise roll, followed by pulling the nose up by 90° (pitch) will result in a dramatic turn to starboard, leaving the glider heading still in a horizontal plane (although on its side) at right angles to its original course. On the other hand, a 90° pitch followed by a 90° roll will leave the glider heading for the stars (until it stalls)! As a result of this importance of sequencing, by convention the sequence of Euler angles is taken as being roll–pitch–yaw. Using these angles, it is in fact easy to construct

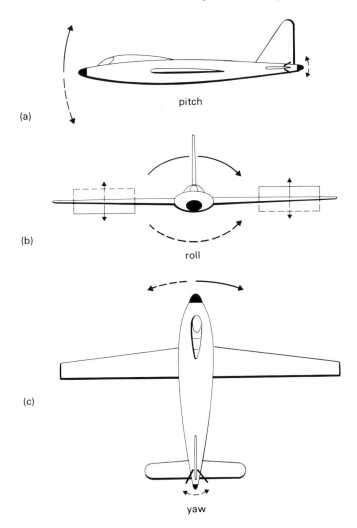

Figure 3.7 Pitch, roll and yaw as applied to a glider

an appropriate 3 × 3 rotation matrix for inclusion in a homogeneous transformation matrix.

Nevertheless, unfortunately, there is in fact no universally accepted convention regarding *which* angles the term 'Euler angles' refers to. Although we have been using the angles corresponding to the sequence roll–pitch–yaw, another commonly employed approach is to measure the angles corresponding to the sequence roll–yaw–roll. If we consider our glider, figure 3.8a, converted for deep-space travel (so that we do not have to worry about gravity), then it can be seen that this new sequence still allows the craft to take up any chosen orientation (for instance pointing directly towards a chosen star). The craft can first be rolled until the plane of the wings and fuselage cuts through the desired star (fiigure 3.8b); the craft can then be yawed round until the nose is pointing towards the star (figure 3.8c); finally, a further roll can align the wings to any required plane (figure 3.8d).

From this it can be seen that roll–pitch–yaw and roll–yaw–roll are

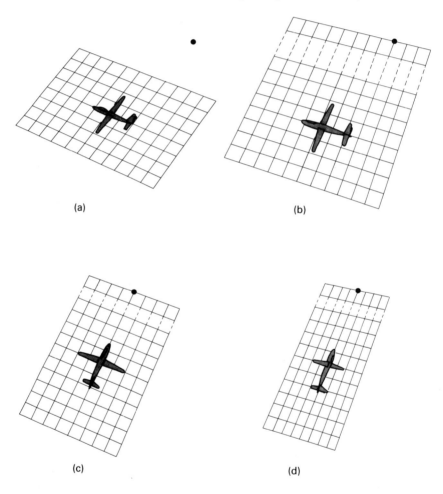

(a)   (b)

(c)   (d)

Figure 3.8 Assuming a required orientation by means of a roll–yaw–roll sequence

both perfectly adequate methods of achieving a desired orientation and, indeed, both can be employed for constructing homogeneous transformation matrices. When considering the actual design of a robot wrist, building three articulations mutually at right angles, as would be needed for a roll–pitch–yaw approach, is frequently more difficult that employing a rotating wrist onto which is attached a yaw–roll arrangement, such as the differential shown in figure 3.2. However, design consideration must also be given to those occasions on which two joint axes align with each other; a rotation about either axis then results in identical motion of the end-effector. In this situation the robot arm has lost one of its degrees of freedom, and is called **degenerate.**

Kinematics

When we consider the movements of a robotic manipulator-arm without any reference to force (that is, the **kinematics** of the arm), we can treat the manipulator as being composed of a series of individual sections, linked together with a particular kind of joint. The joints in such a **serial-link manipulator** can be either **revolute joints** or **prismatic joints**, and the links in effect maintain a fixed relationship between the joints which can then be modelled. A single revolute (**R-type**) joint can only rotate about one axis, and it is the *angle* between the joint in question and the next which varies. With a prismatic (**P-type**) joint, it is the *distance* between the joint and the next that varies. The sequence of joints and links is known as a **kinematic chain.**

Using such a model, it is possible to construct a separate coordinate frame around each joint. Naturally, a consistent method of assigning such frames to each joint must be adopted, and a common approach is the **Denavit–Hartenberg convention** which prescribes a sequence consisting of a rotation, followed by two translations, followed by a further rotation, to bring any one coordinate frame into exact coincidence with the next. From the values of the rotations and translations needed to do this, one can easily derive the homogeneous transformation which describes the relative position and orientation between the two coordinate frames.

Commonly, the homogeneous transformations which describe the relations between adjacent links on a manipulator are called **A matrices.** A given 'A matrix', in other words, is simply a description of the change in orientation and position between that link and the next. Thus, matrix A1 refers to the position and orientation of the first link of the manipulator, while A2 refers to the relative transformation between the coordinate frame of the first link and the second link. As explained earlier in this chapter, it is possible to work along the kinematic chain, multiplying all the relative transformations (A matrices) together, to obtain the absolute coordinates (in the base system) of any particular link. In other words, the position and orientation of the third link, in base coordinates, is obtained by multiplying (using matrix multiplication) A1, A2 and A3 together.

Such products of A matrices are commonly called τ **matrices**, and, for a six link (i.e. six degrees of freedom) manipulator, the absolute location of the end-effector (on the sixth link) would be given by the T matrix:

$$T6 = A1.A2.A3.A4.A5.A6$$

So, knowing the individual A matrices for each link at a given time (in other words a given arm configuration) it is possible to calculate the resultant position of the end-effector (which in practice is frequently the only part of the robot for which position and orientation information is of direct concern). Indeed, it is also possible, after performing such a series of transformations, to extract from the resultant T matrix the orientation of the end-effector in terms of Euler angles. For practical robot work this can be of particular benefit.

Frequently in robotics the most important kinematic calculation is actually to obtain the individual joint positions of a robot arm, given the absolute position of the end-effector (T6). In other words, we know the current position and orientation (the **pose**) of the end-effector; we also know where we want the end-effector to end up. What is required is a list of all the new joint positions necessary to result in the desired new position of the end-effector! The solution can be of vital importance for controlling the arm, yet deriving the form of such a solution for, say, a six-axis arm is a far from trivial task.

There is no algorithm by which the appropriate kinematic equations needed to solve the problem can automatically be derived and, generally, geometric intuition is needed to determine the solution. Nevertheless, once satisfactory kinematic equations have been constructed for a particular arm design, it is then, of course, quite straightforward to compute any required arm configuration by simply substituting into the explicit equations. Naturally, although there is only one end-effector pose, T6, corresponding to a particular set of joint positions, there may be *more* than one possible arm configuration which results in exactly the same end-effector pose. For instance, with a conventional anthropomorphic jointed-arm robot, the same gripper pose might be obtainable

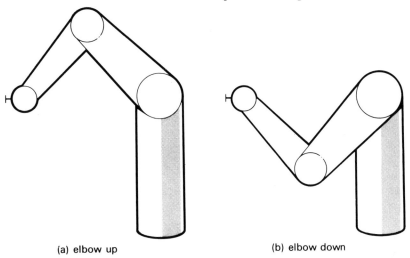

(a) elbow up   (b) elbow down

Figure 3.9 Two possible arm configurations with identical end-effector poses

with the 'elbow' either up (as shown in figure 3.9a) or down (figure 3.9b). Usually, however, it is only in fact necessary for one of these alternatives to be calculated.

Dynamics

So far in this chapter we have only been concerned with the kinematics of a robot arm, without any consideration being given to the forces and **torques** necessary to actually achieve those motions. Nevertheless, with increasingly available fast computing capability, it is becoming feasible to incorporate a model of the **dynamics** of a given arm into the control strategy of that arm. Use of dynamics for control purposes is referred to again in chapter 5.

In practice, multiaxis robot arms represent a highly complex dynamic system, and analysis of these dynamics is consequently a difficult task. A common approach to solving these problems is to use **Lagrangian mechanics**, which is a mathematical technique which enables dynamic equations to be obtained for very complicated systems. (The 'Lagrangian' is basically the difference between the kinetic energy and the potential energy of a system expressed in any convenient coordinate system, not necessarily cartesian coordinates.)

Having developed the Lagrangian for a particular robot manipulator, it is then possible to obtain the appropriate dynamics equations which relate forces and torques applied to the arm to positions, velocities and accelerations of the arm. In other words, given the forces and torques applied, the equations specify the resultant motions of the arm. Of course, thanks to the solutions to the kinematic equations already available, it is not necessary actually to *solve* the dynamic equations (which for all but trivial cases are unsolvable anyway), because *the desired motions of the arm are, in fact, already known* and we simply need to discover what forces and torques must be applied in order to achieve them.

The dynamics equations are obtained in matrix form and are typically extremely complex, comprising several thousand terms. Nevertheless, by analysing which aspects of the equations are, in fact, the most significant, it is easy to simplify the equations down until they are at a level at which they can be computed sufficiently rapidly to make them directly applicable to direct control of the arm. The information which is of particular interest from the simplified equations is the **effective inertias** of each joint of the arm (in other words, the relationship between torque and resultant acceleration at a joint), and the **inertial coupling** between joints. This cross-coupling is, in effect, the relationship between application of a torque at one joint and the resultant accelerations of *other* joints. If the inertial couplings are small, then it is possible to simplify the whole model of the arm by treating it as a series of independent mechanical systems.

In addition to effective inertias and inertial couplings, it is also necessary to determine what additional torques must be applied to joints of the robot arm in order to counteract the effects of gravity. Naturally, in certain positions the weight of the arm may be fighting against the desired motion of the arm; in other postions it may be working with the drive systems; in still other positions it may have no effect whatsoever. There are many other possible effects which may be taken into account, such as friction. In practice however, there are some effects which are small most of the time, and only tend to become significant at high speeds, at which time the accurate positioning of a robot arm is not often of great importance. Consequently, such effects can often safely be disregarded.

Mobile platforms

Throughout this chapter we have only considered robot structures which can loosely be grouped under the heading of 'manipulator arms'. As mobile robots are increasingly becoming commercially viable however, it is only appropriate that some of the different structures employed for such devices should be briefly considered in the concluding section of this chapter. Mobility is dealt with further in chapter 12.

In many mobile robots, of course, there is a 'conventional' robot arm located on top of the mobile platform itself. Such an arm is likely to be identical to those found on a fixed platform, and so will not be considered further here. The structure of the platform, however, is commonly one of those shown in figure 3.10. It can use either legs (figure 3.10a) or else wheels (or tracks) for imparting motion, and the wheels can either be steered (figure 3.10b) or be a special omnidirectional variety (figure 3.10c) which can move in more than one direction relative to the axel.

Legs (as with humans) are structurally very similar to manipulator arms, yet the severe control problems involved in turning corners or traversing rough ground mean that such systems are still only at the research stage. On the other hand, there are many existing practical designs of wheeled or tracked robot platforms. On the whole vehicles incorporating steered wheels require room to turn corners and are difficult to back up, but nevertheless they have been successfully employed in various research devices, and this system is of necessity used when converting such existing vehicles as fork-lift trucks for robotic operation.

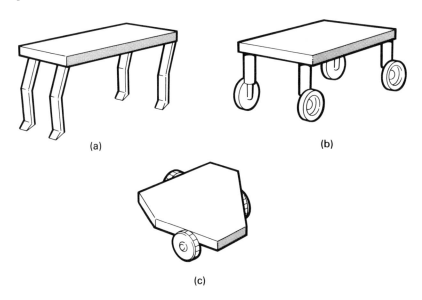

(a)   (b)

(c)

Figure 3.10 Various constructions of mobile-robot platforms

Ackerman steering, as used on cars, is usually considered to be too hard to control in confined spaces to be practical for robotics, and so frequently only one wheel is in fact steered while two others are driven to propel the vehicle. A much more common system however, which approaches the manoeuvrability of a truly omnidirectional system, employs only two (unsteered) independently driven wheels, together with one or more passive 'castors' (provided merely to maintain balance of the platform). If both wheels are driven in the same direction and at the same rate, the platform moves forward in a straight line (figure 3.11a). On the other hand, if the two wheels are driven at the same rate but in *opposite* directions, the platform will rotate on a point midway between the two driven wheels (figure 3.11b).

In this way the platform can be made to travel in a series of straight lines between points, by moving to a point, stopping, rotating till pointing in the next required direction, then moving forward again on the new bearing and so on. Of course, by driving the wheels at *different* rates it is possible to spin the platform about any point on the line between the two wheels (with the wheels rotating in opposite directions – figure 3.11c), or to drive the platform in an arc whose radius depends on the ratio of wheel speeds (with both wheels rotated in the same direction – figure 3.11d). The same properties apply to any tracked vehicles laid out in an equivalent configuration (like a tank).

To minimise the floor space required to turn corners, and to eliminate the control problems associated with steering, some mobile-robot designs incorporate a form of omnidirectional transportation mechanism. Various such mechanisms have been invented, and among the more

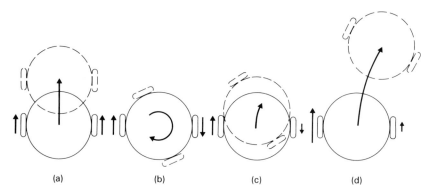

(a)   (b)   (c)   (d)

Figure 3.11 Characteristics of independently driven yet unsteered wheels

revolutionary are some involving specially designed wheels which have rollers mounted on their periphery. In the three-wheel variety, shown in figure 3.12, the undriven rollers are located along the circumference of each wheel, and the wheels themselves are mounted round the platform in the shape of an equilateral triangle. A given wheel can move in any direction by a combination of rotating on its axle like a conventional wheel and rolling on its rollers. Two or three wheels are driven at any one time to achieve the desired direction of travel, and because there

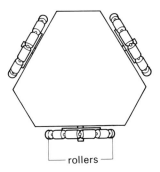

Figure 3.12 Design of three-wheeled omnidirectional platform

are only three points of ground contact, no suspension system is required.

The four-wheel design, shown in figure 3.13, consists of a platform with four conventionally located wheels, but with rollers mounted obliquely around the circumference of each wheel. The platform moves forward and back in a conventional manner, but it can also move sideways by driving each wheel in the opposite direction to its two

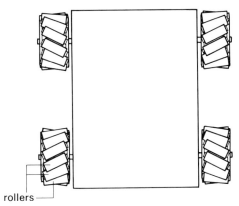

Figure 3.13 Design of four-wheeled omnidirectional platform

neighbours. Similarly, the platform can be made to spin around by simply driving the wheels on one side in the opposite direction to those on the other side. Nevertheless, because of the four-point ground contact it is necessary to employ some form of suspension system to ensure that all four wheels maintain wheel-to-ground contact at all times. In addition, the system suffers from poor efficiency.

As mobile robots are increasingly used in both the factory environment (for such tasks as infrequent loading of widely separated machines) and the nonindustrial setting (such as for commercial greenhouse work), considerations about suitable structures of mobile platforms will become steadily more important. One of the other factors involved in designing such devices is covered in the discussion in the following chapter concerning suitable robot drive systems. A drive system which is suitable for a robot manipulator fixed to the floor may be totally unsuitable when the same manipulator has to carry its own power supply around with it wherever it goes!

4

Power to one's elbow

ROBOT MECHANICS – II
DRIVE SYSTEMS

Hydraulic drives

When the first Unimates were installed over two decades ago, they moved to the accompaniment of the whine and clank of a hydraulic drive mechanism. It seemed a natural approach to take: the controller available was relatively primitive, the required payload for the robot was higher than seemed suitable for electric motors, and manually controlled hydraulic devices of similar kinds had already been in use for decades in such diverse forms as articulated gantries for reaching street-lights and hydraulic attachments for the back of farm tractors.

The great attraction of hydraulic systems is that they can exert large forces (for the size of actuator) by employing high-pressure liquid, usually at pressures around one hundred times that of the atmosphere. Although the equipment needed to create the high pressure originally may be bulky, the fluid can then be directed along suitable narrow piping so as to actually cause movement of a robot joint using only a comparatively small device. In addition, the fluid used is commonly some form of mineral oil, which has the added benefit that the moving parts of the hydraulic system are automatically lubricated. Occasionally however, where fire hazard is an important consideration, the oil must be replaced with a nonflammable alternative such as a solution of glycol in water. For high working temperatures, such as might occur in foundry work, suitable hydraulic fluids are such substances as phosphoric or silicic acid esters.

There are various methods employed in robot designs for converting the high pressure of the hydraulic fluid into useful movement. Basically, the approaches can be split into **hydraulic pistons**, which result in fixed-length linear or rotary movements between given points, or **hydraulic motors** which provide continuous rotary movement. **Rotary vane actuators**, a form of hydraulic piston, are designed to provide a rotation

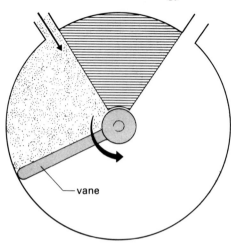

Figure 4.1 Rotary vane actuator

of less than one revolution. They consist, in principle, of one or more vanes (usually only one for robots) projecting radially from a central shaft (shown schematically in figure 4.1) which can be made to rotate inside a casing by forcing fluid into the chamber, so pushing the vane (connected to the shaft) round until it reaches the desired angular position. Motion back again is accomplished by forcing fluid into the other chamber on the opposite side of the vane. **Hydraulic cylinders** (or **hydraulic rams**), on the other hand, provide linear motion (as shown in Figure 4.2) by forcing oil into one end or the other of a cylinder with a plunger in it.

There are various forms of hydraulic motor which can be employed, depending on the cost and performance requirements of the robot. In practice, hydraulic motors have not been common in robots, because the need for a transmission mechanism tends to destroy many of the advantages of using huydraulics in the first place. Use of modern **harmo-**

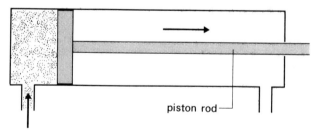

Figure 4.2 Hydraulic cylinder

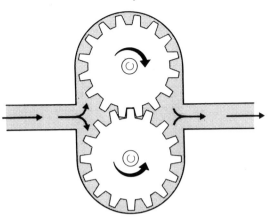

Figure 4.3 Hydraulic 'gear' motor/pump

nic transmissions (discussed later) may alter this in the future. In one design of motor, hydraulic cylinders are attached to a drive wheel to convert their reciprocating motion to rotary motion (as the steam cylinders did on the old steam railway locomotives, or as the pistons do in an internal combustion engine). Alternatively, fluid can be forced into a casing which closely surrounds two intermeshed gears (as shown in figure 4.3). Because the fluid is unable to pass between the gears (where the teeth are enmeshed) it forces its way round near the surface of the casing (in the gaps between the tips of the teeth). In so doing, the two gears are rotated.

In a third form of hydraulic motor, illustrated in figure 4.4, a series of

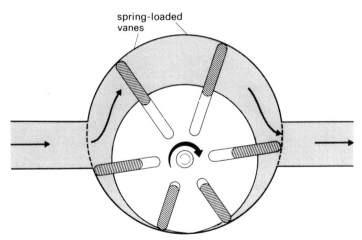

Figure 4.4 Hydraulic 'vane' motor/pump

spring-loaded vanes is located in a rotor on the end of the output shaft. The rotor itself is placed off-centre in a surrounding cylindrical casing. In this way the spring-loaded vanes push out to form a series of chambers but because the rotor is not concentric with the chamber, a given chamber will vary in size as the rotor revolves, reducing to practically nothing where the rotor is closest to the casing. Using this arrangement, oil forced in at one side of the casing will not be able to get past the point where the rotor is closest to the casing, but by pushing the vanes (and rotor) round, the oil can pass through the motor via the larger chambers.

Deciding which form of hydraulic drive to use in a given robot design depends on many factors, but generally the simpler the drive the better. This implies that the simple hydraulic cylinder (which is very reliable and the least exensive option) is the most attractive of all the systems, and indeed it is the most common hydraulic drive; either moving a telescopic joint directly, or providing a rotation by means of acting on a lever. For instance, the shoulder rotation A' of the robot manipulator in figure 4.5 is effected by cylinder A, while the forearm rotation B' is caused by cylinder B.

If superior dynamic performance is required, however, it is necessary to employ one of the designs of hydraulic motor, linked to some form of gearing system. Although this approach tends to be more energy efficient and provides better performance than pistons, it is nevertheless much more expensive, and, being mechanically more complex, is less reliable. Consequently, maintenance costs will be higher.

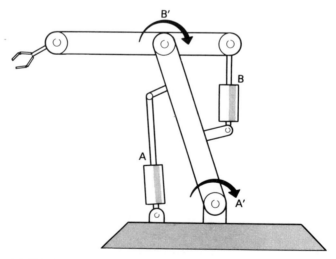

Figure 4.5 Hydraulic arm with rotation effected by hydraulic cylinders

When the operation of the three hydraulic motors mentioned above is considered, it becomes clear that each design can also equally well be employed as a form of hydraulic pump. A piston pump can be formed by driving a rotating wheel (usually by means of an electric motor) which in turn causes linear pistons to suck in and pump out oil (like a bicycle pump). By employing more than one piston, each attached at a different point round the wheel, it is possible to reduce the pulsations of output which would otherwise be produced.

Driving the two gears in the second form of motor described above turns it into a pump. Oil from the input to the casing will become trapped in the gaps between the teeth-troughs and the casing, and will then be pulled round with the rotation of the gears to be eventually expelled at the exit port. Similarly, with the vane motor, driving the rotor round sucks oil into the expanding chambers near the input port but then forces the oil out again as the chambers start to contract near the output port. Any of these three types of pump may be employed to provide the necessary fluid pressure to drive a hydraulic robot. In practice, the piston pump can often produce up to double the pressure of some other varieties, equivalent to the pressure of a few hundred kilograms per square centimetre! Most robots however only employ either a gear pump (usually for lighter duty systems) or a vane pump (for heavier work).

Despite such high pressures, it is not usual for the hydraulic pumps in robots to be able to pump a particularly high *volume* of oil in a given period. Nevertheless, when a telescopic piston is suddenly required to move at full speed through its maximum stroke, then, for a short period, the hydraulic system will require a comparatively large flow of fluid, higher than the pump could produce directly. This is catered for by having the pump store high pressure fluid in a **hydraulic accumulator**. The pump either forces oil into a storage piston which is spring loaded or keeps the oil at the required pressure by means of high-pressure gas acting upon a diaphragm. It is this accumulator which then feeds the remainder of the system.

When the demand for hydraulic fluid is greater than the delivery rate of the pump, then the volume of stored high-pressure fluid in the accumulator will diminish. However, this sudden requirement will not continue for long (the robot might next be required to stand idle for a few seconds) so the pump will soon be able to restore the depleted buffer. In practice, robots such as the Unimate 2000 can supply over three times the flow rate from their accumulators than their hydraulic pumps could directly. This ability of hydraulic systems to 'store' energy and so cope with demand fluctuations, is a distinct advantage over other systems. It is partially offset, however, by the disadvantage that a hydraulic robot will always use the same high pressure of fluid, even when lifting only a few grammes!

In addition to the pump, accumulator and actuators, the remainder of a robot's hydraulic system tends to be complex. It is necessary, for instance, to incorporate several valves both to control the direction and flow of fluid and also to regulate pressure. To avoid the risk of damaging the system, it is necessary to include a **relief system** which will allow fluid to flow back from any actuator, via a relief valve, if (possibly because the arm is attempting to lift too heavy an object) the fluid pressure rises any higher than a preset value, rather like the valve on a pressure cooker. Unlike the pressure cooker, the hydraulic fluid cannot be allowed to escape into the surrounding air, but must instead be conveyed back through piping to a hydraulic fluid tank. The same applies to all the fluid forced out of an operating actuator (forcing fluid into one chamber of a piston, of course, pushes fluid out of the other chamber). So a return-flow pipe network must be provided for the whole system.

Servo-control valves are necessary to control the passage of both high-pressure fluid to an actuator, and also of the low-pressure return-flow from it. Consequently, the electrically activated valve must usually not only be able to shut off or open two pipes simultaneously, but, in order to allow reverse operation of a device, must also be able to 'swap over' the final destination of the two pipes (that is, for example, connect the high-pressure line to either one of the two chambers of a linear actuator). Owing to the complexity needed for proportional control (rather than simply on/off), such selector valves tend to be rather expensive.

In addition, the cost of the valves does not tend to drop with smaller size, and so they can become relatively very expensive in small robots. Other problems with hydraulics (such as cold temperatures increasing the viscosity of the fluid so possibly making a robot 'sluggish') are rapidly making hydraulic robot drives a less attractive option for many applications than the major alternative of electric-drives (covered in the following section).

Recently, however, it has been suggested that **electro-heological (ER) fluids** might eventually provide an ideal solution to this problem for robotic engineers. These so called '**jammy fluids**' consist of a suspension of minute nonmetallic particles (about a thousandth of a millimetre in diameter) in an oil medium. Although under normal conditions the fluids act like conventional liquids, they possess the unusual characteristic that if a high voltage is thrown across them they immediately 'freeze' in the vicinity of the electric field and act very like solids. This rapid change of state can occur (in either direction) within a millisecond.

It is easy to imagine how the properties of such fluids could be used in robotic systems to provide the means for compact, low-cost 'valves'. An optimal arrangement for producing the ER effect has been found to be a 2000 V field across a 1mm gap. To control the flow of ER fluid through a pipe it would simply be necessary to have it flow between two plates fixed parallel to each other and interfaced with the electronic system controlling the robot. When no voltage is applied to the plates the fluid will pass freely. The moment the voltage is switched on, however, the fluid between the plates will solidify and block the flow. By arranging to have two separate such 'valves' situated on either branch of a fork in the hydraulic piping, it would be easy to divert the flow of fluid down either of the two paths, simply by switching the voltage from one set of plates to the other. Nevertheless, such practical systems are still considered to be several years off (if indeed they ever materialise at all), and at present the cost of ER fluids themselves is very high.

Electrical drives

While many people may never have actually had much experience with hydraulic systems, everyone has daily contact with electric motors. Whether they are DC (often battery operated) motors for starting the car or powering a child's toy, or AC (mains operated) motors for turning the spin-dryer or the hands of an electric clock, their capabilities are well known. Nevertheless, there are various distinct types of electric motor which are being used in robotics, together with some new varieties emerging from the research laboratories, and so, before comparing the advantages and disadvantages of electrical drives to hydraulic systems, it is worth first looking in a little detail at what sorts of electric motors are actually most suitable for robotics.

The first commercial electrically driven industrial robot was introduced in 1974 by the Swedish giant corporation ASEA. Traditionally, roboticists have employed DC (direct-current) motors for electrically driven robots, because, not only are powerful versions available, but they are also easily controllable with relatively simple eletronics. Although direct current is needed, batteries are rarely used (for non-mobile robots) but instead the AC supply is 'rectified' into a suitable DC equivalent.

The DC motor simply comprises a set of electromagnets attached to the output shaft, which in turn is situated between the poles of a single large magnet (either permanent or electromagnetic, although usually permanent in robots). The large magnet is designed to 'pull round' an energised electromagnet, so that the two sets of poles are aligned. In fact, each time a given electromagnet is attracted by the large magnet and aligns itself with it, a mechanical switching mechanism on the output shaft changes the electrical connections to the coils so that the large magnet no longer attracts that coil, but attracts the next coil instead. In this way the electromagnets are set spinning round, constantly 'chasing after' the attraction of the large magnet.

Basically, the higher the voltage supplied to the coils of the motor, the faster the motor turns, so providing a very simple method of speed control. Similarly, varying the current controls the torque. Reversing the polarity of the voltage causes the motor to turn in the opposite direction. An explanation of how a controller knows the speed at which a given joint should be moving is provided in the next chapter, but once a value for that desired speed *is* known, it is only necessary to convert the speed value to a proportional voltage level (and then if necessary to amplify that voltage) in order to drive the DC motor directly.

A less well known kind of motor, mentioned in chapter 1, is the **stepper motor**. This type of motor has the ability to rotate at speed in either direction, as well as to stop in any one of a precise number of rotational positions. Some of the more complex examples can actually rotate their output shafts accurately to any one of 200 different angular positions per revolution (that is, less than 2° per step)! It is this sort of motor which is used on certain printers, for instance, to spin round a 'daisy wheel' until the correct character is in position.

The principle of operation of the stepper motor is, in fact, rather the inverse of the DC motor, with a large number of electomagnetic coils surrounding a *permanent* magnet attached to the output shaft, and with the mechanical switching mechanism replaced by an electronic equivalent. In this way, the rotation of the output shaft does not automatically result in any switching of the coil connections which would cause the permanent magnet never to 'catch up with' an attracting coil. Instead, by electronically switching the correct combination of coil polarities, it is possible for the shaft not to be attracted by any coils other than a specified one. Thus the unique absolute shaft position corresponding to that particular switching combination is known.

If the equivalent switching combination is then selected which results in the permanent magnet being attracted only by the coil *next* to the one that previously attracted it, the shaft will rotate through the angular distance of the spacing of the adjacent coils. By subsequently attracting the magnet towards each adjacent coil in turn, it is then possible, under complete control, to spin the shaft at high speed, yet at all times knowing its absolute position. Because of this control of position, it is not usually necessary to check where a stepper motor has stopped, so making it ideal for low cost 'educational robots'. Nevertheless, for larger industrial robots, stepper motors currently tend to be too expensive and offer too little power, as even the largest are usually under one horsepower. The heaviest payloads for stepper-motor driven robots tend to be about 5 kg, as with the 'Locoman' built by Pendar Robotics, UK.

Until recently, AC (alternating current) motors have not been considered suitable for robot design, because of the problems involved in controlling their speeds. In its simplest form the AC motor consists of external electomagnets around a central rotor, but without any form of mechanical switching mechanism for the electromagnets. However, because alternating current (such as the mains electricity supply) is constantly changing polarity (first flowing one way, then the opposite way, several times a second – 50 in the UK, 60 in the USA), it is possible merely to connect the AC supply directly to the electromagnets, and the alternating reversal of direction of current through the coils will perform the same sort of task of changing polarity of the coils as the mechanical switching did on the DC motor; the magnetic field of the coils will appear to 'rotate' (almost as if the coils themselves were being mechanically rotated).

Nevertheless, the speed of rotation of the rotor on the output shaft (which is 'dragged round' by the rotating field) will inevitably be pretty well locked to the frequency with which the alternating current reverses direction. With an AC motor, reducing the voltage will not affect the speed of rotation; the motor will still rotate, locked to the AC frequency, until the voltage is so low that the motor simply stalls. The only satisfactory way to vary the speed of the motor continuously is to vary the AC frequency proportionately – half the frequency will rotate the motor at half the speed.

Until recently, varying the AC frequency (independently for each motor on a robot) was too difficult to be practical. Now, however, due to the development of modern electronic devices capable of handling high power, it is possible to generate varying frequencies of the power required cheaply. Thus each current reversal causes the motor to rotate a certain amount, and the more reversals per second the faster the motor runs. Such motors have the advantage that they can be very powerful, and already manufacturers such as Fanuc have incorporated them into commercially available industrial robots (figure 4.6). It is

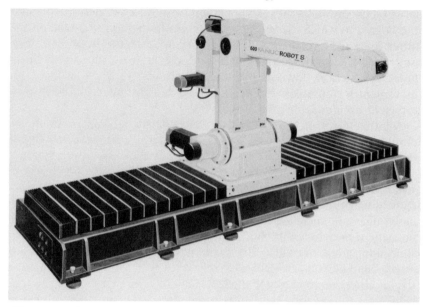

Figure 4.6 Example of an AC servo-driven robot on a track providing limited mobility

widely assumed that low-cost AC drives will eventually slowly displace the less durable DC systems for most robots.

So far, all the electric motors we have discussed have tended to rotate at high speed but only with a small **torque**. As a result, it is necessary to use **transmission** mechanisms with a high gear reduction ratio to transmit power from the motors to the load while increasing the driving torque. Such transmissions are considered later on in this chapter. Recently however, various high torque, low speed motors using **rare-earth** magnetic materials (such as samarium/cobalt) have been developed and are becoming available for industrial use. These **rare-earth motors** are light weight and compact, and because of their performance they can be attached directly to a load without the need for a transmission.

As a result, various research work is being conducted (for instance at Carnegie-Mellon University and MIT, USA) to incorporate rare-earth motors into **direct-drive robots**. These robots have each joint powered directly by a motor without any form of transmission. Naturally, this approach can be problematic because each motor is located at the site of the joint it is powering, so that, for instance, the motor powering the shoulder must also bear the weight of the motors at the elbow and the wrist (conventionally these joints might be powered remotely to reduce the degree to which the motor weights add to the effective load on the other joints). Nevertheless, if such difficulties can be overcome, then direct-drive arms potentially offer excellent features such as no **backlash**, low friction, low **compliance**, high reliability and a reduced requirement for complex maintenance and readjustment. A recent prototype at MIT is capable of speeds of 5 m/s.

Another form of electric motor which might become generally incorporated into robot design (especially for large gantry-type robots) is the linear motor. These motors look radically different from any other conventional design of electric motor, in that instead of being constructed to produce rotary motion they produce straight-line motion. Conceptually they can be thought of as AC motors which have been 'slit' down one side and then 'opened out' flat. By carefully controlling the polarity of the coils of the linear motor, it is possible to make the motor slide along a track in a very controlled manner. Alternatively, the linear motor can be fixed, in which case the 'track' will slide along relative to the motor. Linear motors offer the advantages of requiring neither bearings nor a transmission, so maintenance requirements are also very low. The motors are capable of very rapid acceleration, and can travel at about 1 m/s.

Such linear motors, as well as the rare-earth motors previously mentioned, are as yet far from common even in research robots. But what of the more conventional motors? When is it preferable to use electric drives rather than hydraulics? For first-generation robots, hydraulics tended to be used for robots designed to carry a payload of anything over about 10 kg, despite the fact that early hydraulic robots tended to suffer from rather jerky movements, sometimes leaked and could be unreliable. Lighter-duty robots, usually working in a far less hostile environment, were commonly driven using DC motors, as the high cost of servo-valves precluded the use of hydraulics.

Increasingly, however, despite significant improvements in reliability, hydraulic drives are becoming less fashionable for robotics than in the past, not merely for light robots suitable for tasks such as assembly, but also for the heavier duty applications. Indeed, the world's largest robot, built by Lamberton Robotics, manipulates its 1.5 tonne payload using *electric* DC servos. This shift towards electrical drives may partly be due to the fact that maintenance of *any* robot requires electrical and mechanical expertise, but hydraulic robots require a third skill.

Likewise, electrical drives are usually more energy-efficient than their equivalent hydraulic counterparts, and the overall robot tends to be a lot quieter. Nevertheless, there are occasions when, for example, safety considerations might favour a hydraulic drive rather than electric (which could spark off an explosion in a flammable atmosphere), or when the ability of hydraulic systems to withstand higher mechanical shocks might be important. For much heavy duty work it seems likely that hydraulics will remain the most cost-effective option, at least in the immediate future. It is interesting to note, however, that IBM's 7565 assembly robot, with a maximum payload of only 2.3 kg, is driven *hydraulically*. Also, some **hybrid robots** use hydraulics for the major axes, but electrical drives for the wrist.

For mobile robots the obvious choice is to use electrical drives, both for traction (small mobile robots often use stepper motors) and for any manipulator which the mobile platform carries. The manipulator itself is unlikely to be large (if the platform is to be able to transport it), so electric motors would probably have been used even if it had been nonmobile. The platform, on the other hand, must not only move itself and any manipulator around, but must also carry a power supply with it that will last a reasonable period. At the present time the most suitable form of power supply is a variation of the ordinary lead-acid car battery. Such an arrangement can usually allow the robot to operate for up to a few hours. Although an electrical hydraulic pump could, of course, be run off the battery (so as to power a hydraulic system) the inefficient energy conversion from electrical to hydraulic would severely reduce the period for which the robot could work without recharging. Alternatives to the obvious 'all electrical' approach to mobility are considered after the next section.

Other drives

So far in this chapter we have only considered hydraulic and electrical drives, because it is these two drive types which have, in practice, been incorporated into nearly all robots to date. Nevertheless, it is worth considering briefly the other forms of drive which might be used in their place. Pneumatic drives are an obvious candidate. They are relatively cheap, quiet and reliable, easy to build and maintain, and a leak causes little more than an angry hiss! The systems bear strong similarities to hydraulic versions (with the full range of linear and rotary actuators and motors), yet without all the added complexities of a low-pressure return network (air is merely vented into the environment) and with the advantage that suitable high-pressure air lines are usually plumbed in as standard in a factory. Also, pneumatic components are less accurate and much cheaper than their hydraulic equivalents.

There are some fundamental disadvantages however. Air cannot easily be compressed to anything like the pressures which a liquid can. Most systems use pressures only a few times that of the atmosphere, so, unless very large diameter cylinders are used, arm payloads cannot be high. In addition, air does not naturally lubricate as hydraulic fluid does, and indeed it may contain water vapour which encourages oxidation.

Most damning of all, however, is the fact that it is almost impossible to control the position of a pneumatic cylinder simply. Once the pressure inside a cylinder is high enough to overcome the opposing forces, the piston will tend to start to move and keep going until it hits something! To try to stop it midway in its travel (let alone accurately to, say, 1mm) is very difficult, and usually involves simply applying a brake at the correct moment! Likewise, control of the speed of travel is problematic, and typically requires pulsing of the air input at different rates.

It is possible to spring-load the piston so that the force that must be overcome by the compressed air steadily increases the further up the cylinder the piston is pushed, so that a given air pressure will push the piston to a unique point where the force due to the spring exactly balances the force due to the compressed air. However, in this case the spring is actually fighting against the movement that the robot is trying to make, so much of the (already quite low) energy is spent compressing a spring! Similarly, if the robot then picks up an object, the object's weight is likely either to add to or detract from the effect of the regulatory spring, so all the pressure/distance correlations will no longer apply.

It is generally considered that pneumatics are not really suitable for the speed and accurate control of arm position which is needed in robotics. Apart from anything, if the expense of feedback and control equipment is required anyway, it is usually preferable to pay slightly more for an alternative form of drive which has superior characteristics. Nevertheless, pneumatic drives are ideal for the closely related (though nonrobotic) pick-and-place devices which follow a prescribed sequence of motions for which the limits of travel are determined by physically adjusted end-stops.

In these systems, when the controller dictates that a particular joint should move from one end-stop to another, the appropriate air valve is simply opened and compressed air flows into the required actuator, moving it until it hits the stop. As mentioned in chapter 1, some advanced pneumatic manipulators can automatically switch in and out a variety of end-stops, as does the Electrolux in figure 4.7. It is, of course, possible to provide automatic, continuously variable end-stops (possibly moved electrically) while still employing pneumatics to 'do the hard work'. Such devices should be classed as fully robotic.

Figure 4.7 The Electrolux 'MHU Senior'

Mobility considerations

When the energy employed by a robot using any of the drive systems discussed above is compared with that expended by a human performing the equivalent task, it is found that the robot uses something like *one hundred times* the energy! This is an amazing repudiation of the technological adequacy of our robotic drives in comparison with the 'feeble' flesh and blood of muscle tissue. Clearly a great deal of research will have to be conducted before our robots are as energy efficient as an animal, yet at least we know we are reaching for an attainable goal!

Energy conservation is very worthwhile research, not so much because of the savings incurred on a factory's electricity bill (robots do not use *that* much energy), but because the lower the energy consumption of robot joints then the longer they can be run using a given power source on a mobile robot. As robot mobility becomes increasingly important (for *new* robotic applications, not current ones) it will become vital to increase the length of time between recharges. When a robot is standing idle, recharging its batteries, it is being wasted. In practice it is usually preferable for the robot to simply *replace* (rather than recharge) its power pack with a fully charged one (leaving its old pack to be automatically recharged and then picked up again later). Even so, this option is expensive, and if the robot is out surveying a desert, it might not be very practical either. . .

So what are the short term alternatives? Well, for a start, rapidly spinning flywheels are an efficient method of storing mechanical energy, and have the advantage that they can not only give out energy but, by using spare energy to spin them a bit faster, can actually soak up extra energy for subsequent reuse. In this way it would be possible for a robot, for example, to take advantage of the fact that gravity assists it in lowering a heavy payload. Instead of expending energy having to fight against gravity during the lowering movement (so that the arm does not drop like a stone), the robot could instead convert some of the potential energy of the payload into kinetic energy in the flywheel, to be used later in helping to lift the payload up again. Similarly, a mobile robot moving down hill could convert some of its potential energy (that would normally be wasted using brakes) to be re-used later. Indeed this approach has already been used experimentally in some continental buses designed for hilly regions.

Of course, when large mobile energy requirements occur in everyday life (for instance, to somehow lift up and carry a wife, children, luggage, family dog, radio, heating system and comfortable chairs for several miles along a remote and bumpy track) we do not worry much about energy conservation, instead we simply convert chemical energy into mechanical by burning hydrocarbons, shove all the family, chairs and so on into a metal frame, and call the thing a car. There is not really any reason why the same approach cannot be adopted for mobile robots, at least those designed for outdoor use. If this is done, then, just as with the automatic transmission in a car, use of hydraulic systems becomes perfectly feasible.

Naturally, mobile robots for indoor work would not be popular with human workers if they filled the environment with exhaust fumes. Nevertheless, not only is the need for energy conservation less acute within a building, but if the truly unmanned factory ever became a reality, then the restriction on local atmospheric pollution might cease to apply. Different forms of 'power pack' might, of course be feasible. Instead of burning petroleum-type products, it is perfectly feasible to run an internal combustion engine using hydrogen (many experimental cars in the USA have been so converted) which has the advantage that the only byproduct is water vapour.

Finally, the 'atomic power module' beloved of science-fiction writers, is in reality comparatively simple to manufacture, and has been serious-ly suggested as a method of powering surgically implanted prosthetic artificial hearts. Naturally, it would seem ideal for all forms of mobile robot. Unfortunately, the plutonium used to power such minigenerators is almost unbelievably poisonous. Up till now worries about the possibility of accidental rupture of the protective casing round the plutonium (for instance in an explosion) causing spilling of the lethal contents, has resulted in severe doubts as to the practicality of such a device.

Transmissions

Except in the case of direct-drive robots, each drive unit in a robot requires some form of transmission mechanism. There are very many such mechanisms, none of which are really specific to robotics. Designs can employ chain drives (rather like on a bicycle) or belt drives. Toothed belt drives can be used to drive not just rotary joints, but telescopic joints as well. Gears can be employed, both for slowing down the rapid rotations of motors and also for producing compound movements in the wrist using the differential mechanism discussed in the last chapter.

Rotary movement can be converted into linear motion in a number of ways, most commonly by using either a **rack-and-pinion** system or else a **ballscrew**. The rack-and-pinion approach, illustrated in figure 4.8, involves turning a small gear (or 'pinion') against a straight toothed bar (the 'rack'). This same principle can of course be employed 'in reverse'. A hydraulic piston can move the rack (or even two pistons can be used, one attached to the rack above and another below the piston) so providing rotary motion of the pinion with very high output torque.

When higher precision is needed, or larger forces have to be transmitted, then a ball screw (sometimes called a **lead-screw**) is employed,

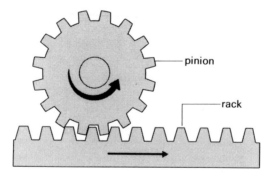

pinion

rack

Figure 4.8 Rack-and-pinion transmission

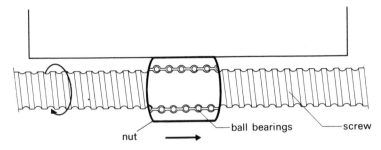

Figure 4.9 Ball-screw transmission

shown in figure 4.9. This mechanism resembles a nut on a long bolt. Turning the accurately machined screw spindle slowly moves the nut, which cannot rotate, along the length of the screw. Anything attached to the nut naturally moves with it. This is the same mechanism that is commonly used, for example, to move the work table of a milling machine. In practice, the nut usually runs on ball bearings, providing a highly efficient transmission (about 90% efficient instead of 30% without the bearings). By, in effect, using two nuts placed together, but pushed apart against the different ridges of the screw, it is even possible to eliminate backlash.

An increasingly common gearing mechanism used in robots is the **harmonic drive**. These systems have been heralded as a major advance in transmission design, because they provide very large speed reductions in one stage (sometimes several hundred to one!) together with high output torques. Nevertheless, because of their novel design, there has been some confusion over how they actually operate, with a few writers claiming that they are essentially the same as conventional **planetary gears** (or **epicyclic gears**) without a 'sun gear'. This is *wrong*.

A harmonic drive consists of three basic components, shown schematically in figure 4.10: a **wave generator**, a **flexspline**, and a **circular spline**. The 'flexspline' is a *flexible* toothed steel ring which on its own could be easily deformed. In practice however it is forced into the shape of an ellipse because of the elliptical 'wave generator' which is placed inside it and acts as a former. This wave generator is specially constructed so that, even if the flexspline is held fixed, the generator can still slide round on the spline's inner surface and so rotate. This is accomplished by fitting a number of ball bearings (in the form of a **ballrace**) onto the surface of the wave generator, so that there is very little friction between the two contacting surfaces of the flexspline and generator. Naturally, because the generator is not circular, as it rotates

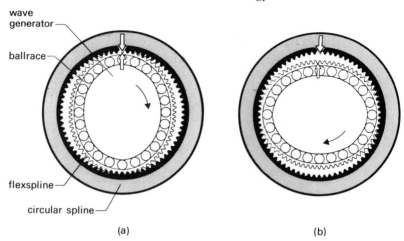

Figure 4.10 Components and operation of a harmonic drive

inside the nonrotating flexspline it will deform the spline which, from the outside, although not rotating, will appear to have two 'bulges' (corresponding to the two points of the ellipse) which move round its circumference.

For a complete harmonic drive, the wave generator and flexspline are then fitted inside the 'circular spline' which comprises an internally toothed solid steel ring. Although the size of the teeth on both the flexspline and the circular spline are the same, there are (say) two less teeth on the flexspline than the total number on the circular spline. As a result, the teeth on the elliptical flexspline only mesh with those of the circular spline at two areas – around the major axis of the ellipse. If the circular spline is held stationary, then as the wave generator is rotated (figure 4.10b) it is these *areas of contact* (not the flexspline itself) which move round with the wave generator, engaging (without any slippage) each tooth of the flexspline in turn with the next tooth of the circular spline.

Of course, because there are two less teeth on the flexspline than the circular spline, once the flexspline has been 'pushed against' the whole of the surface of the circular spline (in other words the wave generator has completed one whole revolution), then the flexspline will in fact end up two teeth away from its original starting point. That is, for each turn of the wave generator the flexspline moves two teeth in the opposite direction! This means that if the wave generator is connected to the input shaft, and the flexspline to the output shaft, then a very high gearing reduction will be obtained. Practical designs of this mechanism are lightweight, compact, have low backlash and are more reliable and efficient than conventional multistage gear systems. As a result, they are becoming very common in electrically driven robots.

Performance

At this stage it is necessary to consider, briefly, the performance of robots. Technical literature supplied by robot manufacturers tends to be rather sparse, but usually lists such specifications as dimensions, number and type of articulations, maximum gripper velocity, maximum payload and so on, which are self explanatory. Nevertheless, when it comes to deciding whether a given robot is capable of (for instance) performing a particular assembly task, then some measure of 'accuracy' is required. It is worth considering what is actually meant by the term: does it refer to how well a robot can automatically follow a straight path between two taught points, or merely how well it can find the two taught points on their own? Does 'accuracy' only refer to the positioning of a robot the same day that it was programmed, or does it take account of any **drift** that may occur over the months the robot is continuously performing the same task? Are specifications taken *after* a warm up period? Do they apply to all axes or just one?

Unfortunately, there is no easy answer! There is no consistently agreed definition for 'accuracy', so in every case it is usually necessary to inquire further to discover what exactly is being referred to. Nevertheless, on the whole roboticists distinguish between **robot accuracy** and **robot repeatability**. *Accuracy* refers to the degree to which the actual position of the robot corresponds to the desired or commanded position. *Repeatability*, on the other hand, refers to the closeness of agreement of repeated movements to the same location, under the same conditions. There is an analogy here with other forms of 'targeting' such as in rifle shooting. A tight cluster on the 'bull' demonstrates high accuracy and high repeatability; a tight cluster, but away from the desired position, shows high repeatability but low accuracy, suggesting that the sights need adjusting; a widely spaced cluster, but basically centred on the bull, indicates low repeatability but high accuracy, suggesting more practice is needed!

In other words, accuracy can be thought of as the error in positioning the first time the robot moves to a point; repeatability is the error (relative to that first position) on subsequent moves to the same point, from the same direction and under the same conditions. Both measure-

D

ments tend to be taken after any vibrations in the robot arm have died down, although this may, of course, assume a pause longer than the robot user might in practice allow . . .

On the whole, repeatability is of more concern than absolute accuracy, and many manufacturers *only* quote repeatability specifications in order to avoid confusion. It is affected by a number of factors, such as rigidity, sensor resolution, payload and speed. Although some larger robots may only be repeatable to 1mm or more, it is common for smaller robots to be repeatable to at least ± 0.1 mm (100μ), that is, about the thickness of a human hair, and some high precision assembly robots are ten times better still! In practice robots do not tend to have the same repeatability throughout their work volume, and so the figure published should be the worst case (usually with the arm fully extended at the maximum reach of the envelope). Nevertheless, the question remains: does 'repeatability' refer to readings taken over one day, one week or one year?

Although there is no consensus, repeatability values are often taken as referring to the robot arm moving (after a suficient warmup period) to a point at the extent of its work envelope, swinging right round (using all axes) to the other side of the envelope, and then moving back to the reference point, continuously, at least for several weeks. The payload and speed used should be the maximum rated values. Some people distinguish between **short-term repeatability** and **long-term repeatability** to differentiate between near-successive repeatability measurements, and those taken over a long period. Naturally, long-term repeatability is likely to be worse than short-term, because any number of different effects (ranging from wear to thermal expansion) can slowly cause the robot arm to drift. It should also be remembered that repeatability measurements are usually taken *under the same conditions,* so may not take account of the fact that removal of a heavy payload may result in **springback** of the robot arm, causing a significant deflection of the end-effector.

Some people subdivide the term 'accuracy' into **playback accuracy** and **dynamic accuracy**. Playback accuracy refers to the difference between a position recorded during a teaching phase, and the actual position response of the robot during the subsequent playback of that point. In fact, because different forms of error are relevant to playback accuracy, depending on whether a point is taught **off-line** or taught using **teaching-by-showing** methods, the teaching method should be specified. Axis misalignment, for example, is relevant to the first method, but automatically compensated for in the second. Dynamic accuracy refers to the degree to which actual *motions* of the robot arm correspond to the desired or commanded motions.

On the whole, the faster one wants a robot to move, the more the accuracy and repeatability suffer. It is not realistic to try to build a robot ten times the size of another, yet still expect it only to exhibit the same absolute errors. This is very much tied up with problems of control (tackled in the next chapter). Similarly, having a robot with joints each of which have high maximum velocities does not automatically imply that the robot will be able to perform a *given* task any faster than another robot with lower maximum velocities.

This is because a robot joint does not instantaneously reach its **slew rate** (the speed at which it moves once it has got up to full speed). Instead, it must accelerate up to the slew rate, maintain that rate for a certain time, and then start to decelerate so that it can stop at the desired location. The rate of acceleration and deceleratation will be dependent on many factors, such as the mass of the arm, the power of the drive and the control strategy, and so it is unlikely to be the same for different models of robot.

If a given joint movement is a small one, then the arm may not have reached its slew rate before it has to decelerate. A robot with a 'snappier' response (though maybe a lower actual slew rate) might in fact be able to perform the same movement more quickly because it actually reached a higher speed more rapidly than the potentially faster alternative (it may even have temporarily reached its slew rate before having to decelerate). This implies that being supplied with values for slew rates may not necessarily be much help in deciding which of two robots will execute a given task more quickly.

There are many techniques for actually testing the specifications of a given robot, but, as an example, to provide the 'flavour' of such work, we will consider an experiment for testing a robot arm both for repeatability and for the amount by which it **overshoots** a point when it approaches it at speed. The technique entails using a special cube containing three inductive sensors mutually at right angles. When the cube is placed into a metal 'envelope', the output from each sensor varies, depending on how far it is away from the surrounding envelope, and this variation is recorded on a chart recorder.

Before a given sensor output can be interpreted as a specific distance, the cube must first be calibrated by placing spacers of known thickness between the cube and the envelope, to see what sensor output they correspond to. Once this has been done, the cube can be attached to the end of the robot arm, and a repeating cycle of programmed movements performed: first approaching the envelope at speed from different angles (to test for overshoot), and then remaining stationary in the envelope for a while (to measure the robot's repeatability and also to discover just how 'stationary' the robot really is).

So much for the 'mechanics' of robots. Yet so far we only have a lifeless mass of linkages and drive systems, still a long way from a robot. Next we must provide a method of controlling this collection of machinery: first at a very low level, and then at a level at which we can actually tell the robot what we want it to do. In the next two chapters we will bring the robot to life!

5

Steam engines and computers
CONTROL SYSTEMS – I
CYBERNETICS

Robots and control

A common descriptive breakdown for an industrial robot is to split it up conceptually into three basic components: the mechanical arm itself, the power units which drive the arm and the control unit. Having dealt with the first two system components in previous chapters, this chapter and the next will deal with the controller, first at a low level which is 'hidden' from the user, and then from the higher level at which the user actually interacts with the system.

At the low level, a controller has several functions to fulfil: it must initiate, control and terminate any motions of the robot arm, moving it to specified points in a specified sequence; it must store all those points, orientations and sequences, for future recall; as well as interfacing with any external sensors and devices which may be connected to the robot. Thus, essentially, the controller, in accordance with a set of given instructions, regulates the flow of energy and other resources in the system, in order to accomplish a given operation. To do this the controller may range from a simple electromechanical system, **fluidic logic** system or hardwired electronic sequencer, right up to a microprocessor or full blown minicomputer. With the low cost of computer hardware it is these days rare to use other than microprocessor-based controllers for any but the simplest nonrobotic pick-and-place devices.

As explained in chapter 1, the only robot-like devices not to make use of servo-control are either small desktop educational robots driven by stepper motors, or limited-sequence pick-and-place devices. The pick-and-place systems tend to need only very simple controllers which merely activate given actuators in the correct sequence. No knowledge about the actual positions of the joints is required as each actuator, once started, keeps moving until it hits an end-stop. More sophisticated limit-switch-control systems employ various limit switch stops at certain predetermined points along an axis, and in these cases the actuators move until the appropriate limit switches are reached.

The most common form of controller for pick-and-place systems is a general-purpose microprocessor-based device known as a **programmable controller** (PC) (sometimes known as a **programmable logic controller** – PLC). Such devices can be used for controlling a far wider variety of systems than just manipulators, but their ability to step sequentially through a specified program with limited feedback capability is ideal for low sophistication robot-like devices.

The remainder of this chapter is concerned with the control of servo-robots.

Feedback

If we consider the old steam engines which inspired the early work on control theory, we can see that the problems with not using some sort of sensing device to confirm the action of a control signal, all arise when there is some external varying influence on the whole system. Say we want the output wheel of the steam engine to run at a constant speed. So long as the load on the wheel remains the same, we could just set the lever which regulates the power of the engine to a given position, and once up to speed the engine would probably indeed keep going at the same rate. If we increased the power the engine would settle down at a faster rate. We could possibly even mark positions next to the regulator lever which corresponded to different absolute speeds. But, what if the load on the engine keeps varying, or the steam pressure changes as the boiler heats up? Obviously, all the carefully marked 'absolute speeds' will become meaningless.

In these circumstances, when there is a **disturbance** or **secondary input** to the system, **open-loop** control, without feedback, breaks down. James Watt solved this very real problem by means of his 'governor', shown in figure 5.1, which consists of two heavy spheres hung from a shaft which rotates at the same speed as the engine. The faster the engine goes, the more the two spheres are flung out by centrifugal force, and as they do so they pull up a sleeve connected to a lever controlling the power going to the engine. By arranging the system so that the faster the engine, the more the spheres are thrown out and the more the power-supply level is turned *down* (or conversely, the slower the engine, the less the spheres move out, so the more the steam regulator is turned *up*), it is possible for the engine to compensate automatically for any changes in load by increasing or decreasing the power supply as required to maintain a specified speed.

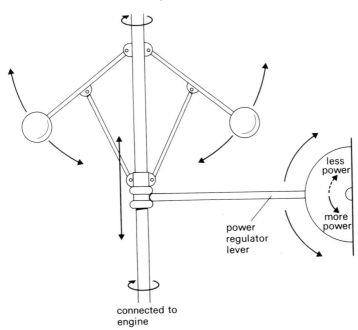

Figure 5.1 Principle of the Watt governor

Such a **closed-loop** control system, characterised by the use of **feed-back**, can be represented schematically as in figure 5.2 All servo-mechanisms exhibit the same kind of properties, whether they are purely mechanical devices like the Watt governor or electronic systems as used in robots. Use is made of some kind of error-detector which returns a signal that is proportional to the difference between the *required* value of something and the *actual* value for it (which is monitored and 'fed back' into the system, so forming a closed 'loop' in the conceptual control process). When feedback is not used, the control loop is effectively left 'open', and so the response and accuracy of the system suffer.

A generalised simple feedback-control system as could be used on each joint of a robot is shown in figure 5.3 A **transducer** senses a required condition (such as position or velocity) and transforms the sensed information into a form suitable for the servo-system (for instance, the voltage put out by the transducer might be directly proportional to the position of the joint). This feedback signal is then subtracted

Robotics Technology

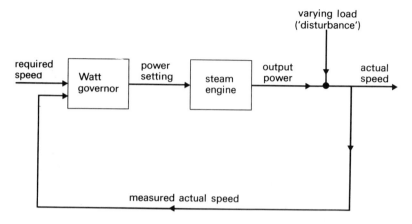

Figure 5.2 Representation of the feedback process employed by the Watt governor

from the input signal (the required position), so producing an **error signal** which, after being amplified to a sufficient power, actually drives the robot actuator system. If the input and feedback signals are the same, then the error signal is zero, so the condition of the system does not change. If however the two signals are not the same, then the amplified error signal will drive the joint in such a way as to reduce the error signal to zero again. Practical systems always employ such **negative feedback**; any system which used **positive feedback** in which an error signal caused the system to produce an even larger error signal would of course be unsuitable

Owing to the very nature of feedback control, there can be a tendency for a system to be unstable, and to oscillate about a desired value rather

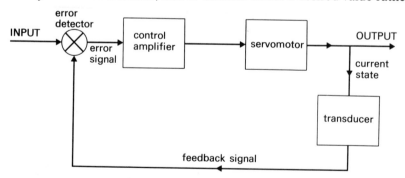

Figure 5.3 Generalised simple feedback-control system

than actually achieve it. Such an **underdamped** system response is shown in Figure 5.4a, where the system reaches but overshoots the desired 'steady state' and proceeds to oscillate about the point with ever decreasing amplitude, **hunting** for the correct state. The opposite problem occurs in **overdamped** systems, illustrated in figure 5.4b, where the system only reaches the required steady state after a long time delay. **Critical damping**, figure 5.4c, exhibits the fastest approach to a steady state without any oscillation, but is nevertheless still rather slow, so in practical systems damping is usually chosen to result in a slight, but tolerable, overshoot (as shown in figure 5.4d) which provides a significantly faster system response.

So far the feedback devices which have been discussed have employed **analog feedback**, where the system operates by the comparison of the different magnitudes of signals. In robotic systems, however, a different type of feedback is commonly used, termed **digital feedback**, where all

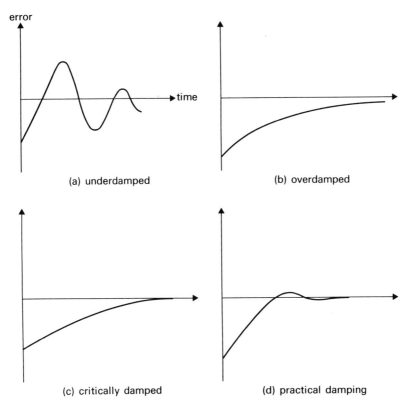

(a) underdamped

(b) overdamped

(c) critically damped

(d) practical damping

Figure 5.4 Different forms of system response

signals are in the form of binary (on/off) pulses. The reasons for this are very similar to the reasons why **digital computers**, as opposed to **analog computers**, are now almost universal: accuracy, low cost, tolerance to electrical noise, and so on. So, whereas with analog feedback, position information might be represented by the magnitude of a voltage, with digital feedback, position data would be conveyed by an equivalent stream of discrete electrical pulses, and a digital comparator would compare the count of feedback pulses with the original value and output a corresponding pulsed error signal.

An example of such a system is shown in figure 5.5. This represents a more sophisticated feedback arrangement than before, as this system includes both position and velocity control. In many robot systems velocity control is essential if high levels of accuracy are to be maintained for final positioning. In the example shown, the servos are driven by an analog voltage, so a **digital-to-analog converter** (DAC) must be incorporated to change the pulsed error signal into an analog voltage before it is amplified prior to subsequent processing. Unlike the position transducer which outputs a digital signal, the velocity transducer(**tachometer**) here is shown as analog, and its signal, proportional to the speed of the joint's motion, is fed back to effectively modify the existing output from the first control amplifier.

By careful design of the two interconnected feedback circuits it is possible to obtain the system response desired. Even without a velocity feedback loop, the positional feedback system on its own will in fact provide a limited velocity control, for the greater the error signal is (corresponding to a large distance between the desired position and actual position) the greater will be the amplified voltage driving the

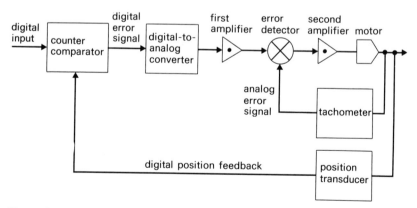

Figure 5.5 System incorporating both position and velocity feedback

power system of the joint. Thus, as the arm lags more and more behind the command signal, so the motor driving the joint will speed up automatically. However, such velocity control exhibits a relatively slow response, and its characteristics are inherent in the positional feedback control. Incorporation of a separate mediating velocity-feedback loop permits rapid correction for velocity changes, and can have a stabilising effect on the overall servo system.

The choice of each amplifier in the multifeedback system has a profound effect on the response of the system. If, for example, the gain of the second amplifier in figure 5.5 is high, then a very small error signal from the velocity feedback loop can produce large changes in the signal to the motor, and correspondingly large variances in the motor speed. Similarly, if the gain of the first amplifier is too high, then the natural 'damping' in the system may be overcome, and the system will suffer from sustained oscillation about the required position. In practice, there tends to be a balance to be maintained between high positional accuracy, rapid response and low oscillation.

When considering the actual physical measurement of position or velocity by an appropriate transducer, there are fundamentally two different approaches that can be taken. With **indirect feedback** the transducer measures the actual motion of the servomotor. Although this may correspond directly to the actual movement of the robot arm (which is the thing the system is actually interested in), it is quite possible that problems such as **backlash**, where the mechanical drive components move a small amount before the elements actually transmit any force, will result in the arm not moving in quite the same way as the motor. As a result, the most accurate method of measurement involves using **direct feedback**, where the transducer measures not the movement of the motor but the actual resultant movement of the joint. Although it is more accurate, direct feedback is usually more costly to implement, and consequently, many closed-loop configurations in fact employ indirect feedback.

Internal-state transducers

Transducers are devices which in effect convert information from one form into another, for instance responding to a physical stimulus to produce an electrical signal. For robotics systems, appropriate transducers for measuring a required quantity tend to be selected for both compatibility with other components in the loop and suitability for the environment in which the robot is to work. In addition to measuring position information, many robots also require information on velocity, direction of movement and sometimes even acceleration.

There are many common approaches to categorising transducers. Grouping may be by the variable being measured; by whether you have to supply energy (as in a potentiometer), **passive**, or whether the energy comes from the system being measured, **active**; by the physical form of the transducer, **rotary** or **linear**; or by the measurement technique employed, such as digital or analog, or **absolute** or **incremental**.

Linear transducers, suitable for example for cartesian-type robots, contain a scale with very fine gradations on it. Each time an optical or magnetic reading head passes over and senses a line, the head emits a pulse, and by counting the pulses it is possible to measure the total displacement of the transducer head. In some systems, in order to measure extremely fine displacements, special gratings are constructed in which a transparent window with fine rulings on it is passed over a long scale with identically spaced rulings. By arranging that the lines on the window are at a sight angle to the lines it is passing over, a **Moiré fringe** interference pattern is built up and creates an illusion of dark bands which move a large distance for a very small displacement of the window. It is these 'bands' which are then counted optically, allowing resolutions of up to 1μ (a thousandth of a millimetre).

With these linear incremental encoders it is necessary to indicate the direction of travel, as the form of the pulses is independent of which way the head is travelling. Various methods can be used to solve this problem, and these usually take the form of placing two or more sensors slightly out of phase with each other with respect to the spacing of the rulings. This results in a unique sequence of pulses from the two sensors dependent on the direction of travel.

Upon occasion it is useful to use an absolute linear encoder, which, although much more expensive than an incremental version, outputs a unique binary code for each different point along the whole travel of the transducer. Such devices carry a special pattern on the scale, which is then read by photoelectric, magnetic or electrical conducting systems. In the simplified example shown in figure 5.6a, a bank of four sensors would be required to distinguish the 16 separate positions. Although the binary code shown seems the obvious pattern to use, problems may arise when the sensors are scanned along the scale, because there will be positions at which adjacent patterns may be read simultaneously, so producing erroneous values.

For instance, if the sensors in figure 5.6a were slightly to the left of where they are shown, the pattern might be interpreted as 0111 (rather than 0011 or 0100). This problem can be solved by using **window code** in

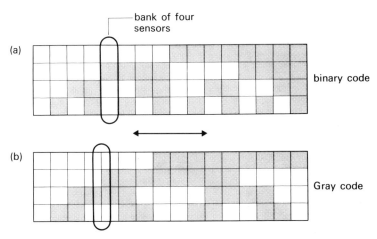

Figure 5.6 Absolute linear encoders employing different pattern sequences

which an additional track is provided of small 'windows' aligned with each pattern, and a reading can only be taken when the sensors are centred over one of these windows (and so not over a pattern change). Alternatively, a different pattern sequence known as **Gray code** can be used, as in figure 5.6b. Because only one bit of the pattern ever changes between consecutive positions, there is never any ambiguity.

All the forms of the above linear transducers also occur in rotary designs, which are popular because of their compact size and flexibility of usage, and which can distinguish up to a few thousand different positions per revolution. Nevertheless, both linear and rotary types tend to be fragile, and vibration may cause innaccurate readings, as may electrical noise. In addition to such digital encoders, however there are two forms of analog rotary transducers known as **resolvers** and **synchros**. Both forms work by feeding alternating currents to fixed coils and measuring the resultant current induced in a mobile coil. Such systems are very robust and can be used in electrically noisy environments, and yet can provide a measure of input shaft rotation accurate to a few tenths of a degree.

To determine the absolute position of a robot which uses incremental encoders, on being first turned on, the robot must be **calibrated** by moving it to a known position from which all subsequent motions are measured. Some manufacturers which take advantage of the simplicity and lower cost of incremental (as opposed to absolute) encoders, never-theless arrange for a recognisable **indexing point** to be passed for every few hundred pulses of the encoder. By using an independent, low accuracy, absolute transducer (such as a potentiometer) in tandem with the incremental encoder, it is possible to assign absolute values to the many 'index points' and to be able to distinguish between them using the cheap absolute transducer. Calibration of this kind of system merely involves passing through the nearest index point.

Point-to-point control

There are fundamentally three different ways of controlling point-to-point motion in a robot, but each makes use of independant position servos for each joint, and for each, the actual path followed by the robot arm between taught points is not of concern. As a result, the robot controller does not need to employ a full kinematic model. The simplest form of control is **sequential joint control** in which one joint at a time is activated, while all the others remain stationary. The resulting path which the end-effector of the arm follows is likely to zig-zag across the work envelope, and consequently the time taken to move from one taught point to another tends to be far longer than necessary. Nevertheless, sequential control requires the simplest of control structures, and may also suit highly modular systems where individual joints may subsequently be replaced by multijoint manipulators.

With **uncoordinated joint control**, although all joints move at once, the fact that one joint has completed a given fraction of its travel does not imply that the other joints have covered a similar fraction of theirs. Each joint moves through its required path in its own time, and then waits for all the others to finish. Owing to the inherent lack of coordination between the different axes, it is difficult to predict the path and velocity of the end effector as it travels between taught points. With **terminally-coordinated joint control**, however, the individual joint motions are timed so that they all start and stop together, making it the most convenient, although the most expensive, of the three methods.

Continuous-path control

When the form of the actual path followed by a robot end-effector between two points is of primary importance, it is necessary for the controller to employ a kinematic model if the intermediate movements are to be interpolated, not all taught. It must then simultaneously move each axis the minimum distance necessary to reach the next intermediate point in space required to keep the end effector on a controlled predicatable path. Such sophistication usually requires a mini- or 16 bit microcomputer as the controller. In addition to a robot following a path of closely spaced points that it has actually been taught, it will then be possible to request that **linear interpolation** be performed between only two widely spaced points (or even 'fitted' through more than two points), resulting in an (untaught) straight-line motion. With some controllers, **circular interpolation** is also possible. For welding robots it is possible to superimpose a **weaving pattern** over the basic path.

There are various approaches to continuous-path control, each using different amounts of information about the actual path to be followed. The simplest method is the straightforward servo-control approach, which makes no use of any knowledge about where the path goes in the future. Although details of the path may be stored in the robot's memory, all the controller refers to when driving the arm motors is the error signal indicating the difference between where the arm actually is, and where the next intermediate point is that it is heading for. It is this method of continuous-path control which is most common in present industrial robots.

More advanced systems make use of such approaches as **preview control** (or **feed-forward**). This method takes into account the way the path changes immediately in front of the end-effector's current position. A still more advanced approach employs **path planning** (or **trajectory calculation**), in which knowledge of the whole path to be followed is incorporated into a mathematical model of the robot arm and the load it is carrying, and a detailed acceleration profile for each joint is computed, together with predictions for the required motor control needed to make the arm follow the desired path. Such a dynamics-based approach, as touched on in chapter 3, can incorporate the effects of forces such as gravity, inertia, damping and friction, and it permits highly accurate movements at speeds which would otherwise be impossible.

Additional sophistication

Use of transformations allows translations and rotations to be easily performed, not just of points but even whole programs. Nevertheless, the complexity of the computation involved in all the transformations can sometimes result in slowness or inaccuracy. Continuous-path control with such systems involves applying an interpolating function to the transformed **world-coordinate** locations, and then, in real time, rapidly transforming them into joint positions which the robot can 'understand'. This must be continuously kept up so that the motor servos have a constant stream of new intermediate points to head for.

A particularly useful application is to provide a **full tracking** capability: there are many occasions in which the task performed by a robot must be synchronised with the motion of a conveyor belt (the arm might be spray painting or welding a passing car chassis). Conventionally the robot would be taught its task with the object moving in front of it, and so long as the synchronisation of robot and object in the future can always be maintained, and so long as the conveyor speed never varies, this method will be perfectly adequate to allow the robot to perform its task on other objects moving in exactly the same way as the first. However, in cases where constant uninterrupted conveyor speed cannot be guaranteed, it is possible to teach the robot with the object stationary, and for the robot then to use coordinate transformation to perform its task relative to the object, and by monitoring the conveyor speed, to be independent of the motion of the transfer line.

6

Do as you're told

CONTROL SYSTEMS – II
PROGRAMMING AND SOFTWARE

Types of hardware

Robots may be getting inherently smarter, but this is largely because humans are getting a lot smarter at programming them! However sophisticated the mechanical linkages and drives of a robot are, and whatever range of sensors is attached, the ultimate performance of the system is governed by the capabilities of the programs controlling it.

With the early robot control systems, the internal process control program was specified by the selection and wiring of relays (and later simple semiconductor logic components). Any changes to the 'program' involved rebuilding the **hardwired** controller. Consequently, not only was 'programming' a very slow process, but the range of possible instructions was very limited. 'Memories' consisted merely of such devices as disc or drum cams, **plugboards**, punched tape, potentiometers or **fluidic** systems, all of very limited capacity. Truly flexible programming had to wait for the introduction of cheap computing power.

Modern control systems (figure 6.1) employ microcomputers (or even their more sophisticated cousins, minicomputers). This immediately results in potentially far more sophisticated programming capabilities, with greater flexibility, yet simplicity of operation, and with far greater ease of reprogramming. The increased computing power permits not only interfacing with sensors, real-time interpolation and moving frames of reference, but also the use of 'libraries' of commonly needed programs.

Memory is frequently split up into fast **primary store** and slower (cheaper) **secondary store** or **backing store**. Primary store often comes in the form of semiconductor devices which although fast and relatively cheap are **volatile** and so lose their contents when the power is turned off. In industrial-type environments, very short power black-outs (or **brown-outs**) may be frequent enough to be a serious problem, and consequently such semiconductor memories are often provided with

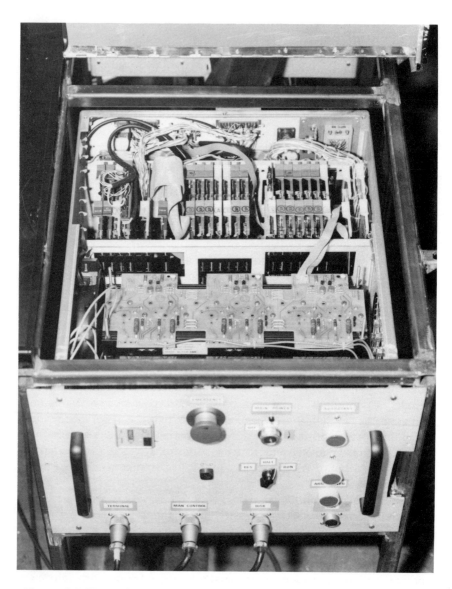

Figure 6.1 Example of a typical modern robot controller

battery back-up systems which can at least maintain the contents in store during a short power cut. Longer term backup and bulk storage of several programs is accomplished using a secondary-storage device such as a floppy disk or tape unit.

In the late 1970s, **dynamic** semiconductor stores became available which, in contrast to the more conventional **static** memory, require their contents to be 'refreshed' several times a second. Although much cheaper and more compact than static memories, dynamic store is more susceptible to electrical 'noise' and consequently, although common in many microcomputers, has not often been used in industrial equipment.

Some older systems make use of **core store**, consisting of a matrix of minute ferrite rings which have the property of not losing the information stored in them when their operating power supply is disconnected. Such devices which hold their information securely are termed **nonvolatile** or **permanent**. A modern equivalent to core store is the so-called **bubble memory** which stores data as microscopic magnetic 'bubbles'. Such memories are not only non-volatile, but can also be manufactured to hold such large quantities of data that a backup storage medium may not be required.

It is possible to use one fast processor to control several different systems (such as different joints in a robot arm) all at the same time. Such **pseudoparallel processing** is accomplished by having the single processor switch between the several systems so quickly that each system appears to have its own processor. With the decreased cost of micro-processors however, it is now common to actually have several processors, for instance a separate processor for each joint of a robot, with possibly an additional processor to 'supervise' them all. Such **multiprocessor systems** can be architecturally configured in a variety of ways, differing by the degrees of autonomy with which the individual processors operate; possibly sharing common memory in a 'tightly linked' network, or in a 'loosely linked' system acting as quite independent microcomputers in their own right. A common configuration is some form of hierarchy, with successive levels of processors dealing with more and more global problems, similar to a traditional human management structure.

Interfacing the robot's computer with the outside world, whether to internal-state transducers or to external **peripherals**, is accomplished via input and output **ports**. More and more, such interfaces are themselves configured by means of internal software, and indeed, it is quite possible for many interface components such as electronic filters, linearising devices and digital-to-analog and analog-to-digital converters, which are traditionally built as physical hardware, to be replaced solely by a software program which when run on the existing microprocessor emulates the actions that the physical devices would otherwise perform! Although there is a consequent speed penalty, it is frequently acceptable (a simple programmed ADC can still recognise frequencies of over 1,000Hz – cycles per second), and the advantages in simplicity, cost, and inherent flexibility are obvious. In fact, there is a trend towards writing more and more of the tasks previously consigned to dedicated hardware as software to run on a processor that would be required anyway. As this trend continues, the delineation between hardware and software increasingly becomes less distinct, and the role of factory programming of specialized software (as distinct from on-site programming of robots) becomes paramount.

Types of programming

Programming a robot to perform a required task may involve a totally different type of 'programming' to that used, for example, when sitting in front of a personal computer. Although it is conceptually relatively straightforward to type in at a terminal a sequence of point coordinates and then have a robot arm proceed through those specified points (which might be an ideal way of programming a **palletising** routine), there would be tremendous problems in trying to use the same programming method to teach a complex paint-spraying operation currently accomplished by a human.

Limited sequence pick-and-place devices do not, of course tend to have any form of true programming ability – they are physically reconfigured instead, which prevents them from being true robots. Nevertheless, such alteration is sometimes referred to as **manual programming**, and involves establishing a new 'logic' sequence by possibly connecting air tubes in a particular way, making **plugboard** connections, or positioning limit switches and mechanical stops. The resultant 'programs' are limited to a very few steps.

True programming of robots can be split into two fundamentally different approaches. Robots can be taught either by **textual programming**, sitting in front of a computer terminal, or by **teaching-by-showing** methods, which are ideally suited to such tasks as the paint-spraying operation mentioned above. Textual programming is a major subject and is dealt with further later in this chapter. 'Teaching-by-showing' methods involve a human somehow physically moving the robot arm through the sequence of positions which are later to be automatically replicated. Such techniques are particularly suitable for continuous-path

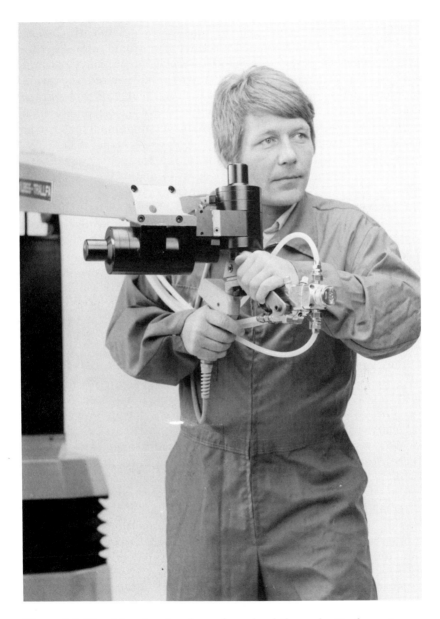

Figure 6.2 Teaching-by-showing using a lead-through attachment

tasks, such as spray surface-coating. There are two popular forms of 'teaching by showing': **lead-through programming** and **walk-through programming**.

Lead-through programming requires some form of lead-through aid in addition to the usual manipulator and controller. Such aids consist either of a special attachment to the wrist of the robot arm, or else a quite separate special **teaching arm**. To teach the robot a task using this programming technique involves the operator in first setting various switches and buttons on the controller to select the correct program number for the operation, the sampling rate at which points are to be recorded in memory, any synchronisation requirements with such devices as conveyors, and finally an interlock which allows the operator to take over manual control of the robot arm. He then manually leads the arm through the required task by going through the complete work cycle.

If the operator is using a lead-through attachment to the robot wrist, as in figure 6.2, then the robot arm itself will be moved by the operator. To achieve this the arm must usually be counterbalanced, and if hydraulic, it must have its pressure reduced so that the operator is not

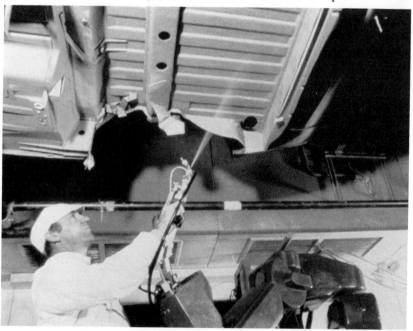

Figure 6.3 Teaching-by-showing using a separate lightweight 'teaching arm'

'fighting against' the robot's normal drive system. If he was teaching something like a paint-spraying task, the programmer would actually paint a sample workpiece using a spray gun as the robot end-effector, and in so doing would move all the robot joints which, by being constantly monitored and their movements recorded, could subsequently reproduce the human's motions. Additional actions such as turning the spray gun on and off are also recorded.

In certain cases the opposing forces which the programmer must overcome in moving the arm (particularly at the boundaries of the work envelope) make subtle motions impossible. In such cases a separate teaching arm, consisting of a very light framework with an identical kinematic structure to the main arm, is sometimes used as a **teleoperator**, as shown in figure 6.3. The great advantage of such lead-through methods is the ease of programming, which can be accomplished by a human, skilled in the process being programmed but with no knowledge of robot languages. However, because paths are recorded by storing a large number of points, any editing would be bound to cause discontinuities, so the only effective way to edit a path using this system is to reprogram it.

In contrast to lead-through, 'walk-through' systems are largely used for point-to-point control and make use of a **teach pendant** like those in figure 6.4. Such portable pendants act as 'remote control' devices with

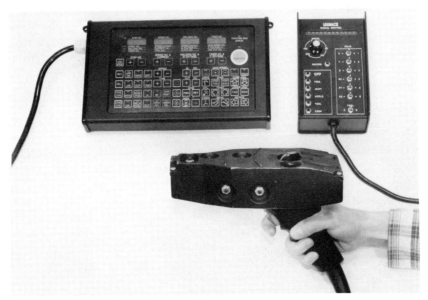

Figure 6.4 Different examples of teach pendants

which a programmer can drive a robot arm to different positions. At its simplest a pendant consists of a switch to drive each joint of the robot, together with a 'teach button' which is pressed once the programmer has steered the arm to a required location (using the other switches) and wishes to have the point recorded for subsequent replay. More sophisticated pendants may include certain function keys for time delays, manual or playback speed control, or outputs and waits for inputs, together with such things as an emergency-stop 'panic button'. Control systems with substantial computer power may allow selection of one of several 'coordinate systems' (such as cartesian or cylindrical) while in the teaching mode, so that by using only one switch the programmer can, for instance, move the end effector parallel to the X-axis, even though that involves the simultaneous movement of several joints.

Such walk-through systems record only the end-points of paths, and the form of the actual intermediate path followed during playback will depend on the form of interpolation used (if any), as discussed in the last chapter. Special library routines can be executed which allow for **logical branching** in the program. Editing of programs can be quite sophisticated, allowing easy insertion, modification or deletion of individual steps, and it is possible to test a program by getting the arm to move through its sequence of points step by step. In addition, it is possible for the controller to support the use of a wide variety of functions such as time delays or output signals, and because only discrete end-points are stored there is room in the memory for large programs. Nevertheless, the walk-through technique tends to limit a programmer to relatively simply structured programs with limited branching capability, taught **online**, and usually with no easy means of documentation.

Types of software

Construction of complex programs for the operation of robots involves writing true software, as opposed to merely programming the robot online by means of teaching-by-showing (although, of course, teaching-by-showing does in fact automatically construct simple software which is subsequently run by the robot). However, before dealing with such textual programming, it is worth considering the wide variety of software used in robotics that is not usually written by the user.

Robotics software can be thought of as ranging from 'robot software', which is necessarily used within robots, through 'robot-related software', which helps to expand the sophistication demonstrable by them, to 'robotics-related software', which covers those areas relevant to both

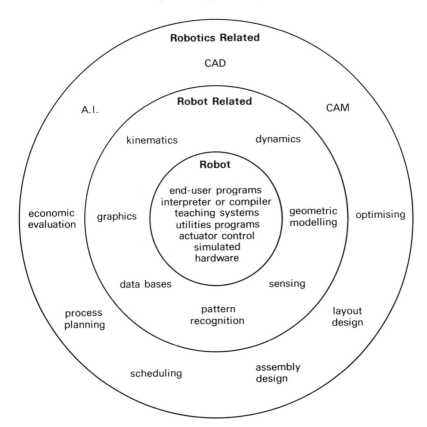

Figure 6.5 Different forms of robotics software

robot systems and others. As with all such classifications, it is not always clear into which group a given type of robotics software should be placed. Nevertheless, one such grouping might look as in figure 6.5.

In the 'central core' of robot software come not only the programs written by the end user, but also the software of the **interpreter** (or sometimes even a **compiler**) which is needed to interpret and translate the user-programs into a form which the controller can actually understand. There is also the controller software, which allows the various teaching systems to function, the utilities programs such as editors, and most importantly the actual control software which responds to feedback signals for each articulator in the arm. In addition, as mentioned in the first section of this chapter, there may be software to simulate hardware components such as analog-to-digital converters.

Robot-related software can come in many forms. Robot systems may attempt to display information using graphics systems that require specialized software to run them, and sophisticated controllers may try to improve robot performance by using kinematic (or even dynamic) models of the arm which have to be able to react in **realtime**, so requiring significant computing power. Likewise, as second-generation robots make more use of sensory information such as vision and **taction**, so the software involved in interpreting the information obtained also increases.

Many sensory devices, such as video cameras and sensor arrays, have the potential to provide so much data that it is infeasible for any central computer to process it all. As a result a **preprocessor** with its own software is commonly employed to simplify the data before it is passed on to the main processor, which may be some distance away. Pattern recognition and other interpretation of sensory data, is frequently far too complicated for the special software required to be run on the robot's computer system, and consequently this additional software is usually run on a form of computer architecture which has been specially selected for its suitability for such **algorithms**.

As robot models of the 'real world' expand, so the collated data must be stored in a specially structured 'electronic filing system'. Such **databases** not only store the required information, but employ sophisticated software to allow retrieval of selected sets of the data. Improved methods for constructing such database-management software are still being developed.

Of course, robotics-related software is drawn from a wide variety of disciplines. The combining of **computer aided design** (CAD) systems with **computer-aided manufacture** (CAM) (to form CADCAM systems) provides the potential for product designs created and stored with the aid of computers to be manufactured automatically from that same information held in the drafting computer. The complexity of the software involved in both systems, together with that needed to interface between the two, is understandably extremely high.

Closely related problems involve attempts to optimise such areas as: layout design for just robot cells or possibly for whole factories; assembly design, ranging from design of the products for ease of robotic assembly to design of the best sequence in which to accomplish the assembly; as well as methods of scheduling and process planning. All the above must somehow be optimised according to economic and other criteria. Again, the software required is very complex, and much work remains to be done.

Finally, an area which seems destined to make a substantial contribution to robotics is **artificial intelligence** (AI). The form of software used in AI systems tends to be radically different from that used in more traditional computing, and ideally requires a different form of hardware architecture to run it on efficiently. Nevertheless, as such new hardware becomes available, so AI programs are likely to alter substantially the ways in which end-users program their robots. Much research is being conducted into AI, and the nature and implications of this vitally important and exciting work are dealt with further in the last two chapters of this book.

User-programming with software

The software in a generalised robot configuration can be thought of as interfacing with two major parts of the system in addition to the programmer: the 'mechanics' which includes the robot manipulator, gripper and any other such components, and the 'sensing devices', which provide the robot with information concerning its immediate environment. This is represented diagrammatically in figure 6.6.

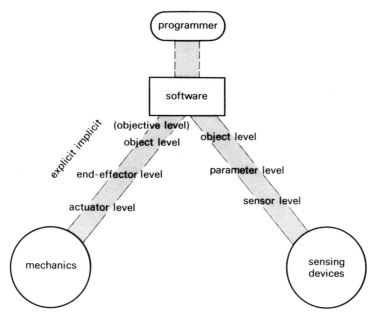

Figure 6.6 The various levels of abstraction of user-software in a robotic system

The software for both these major interfaces can be categorised into three broad levels, dependent on the degree of abstraction they exhibit to the physical manipulator or sensors. At the lowest, least abstracted level of the 'mechanics software', the programs are such that they focus on the workings of the joints and actuators of the mechanical system. Such programming requires for each movement, substantial detail about individual actuator operations, joint positions and so on. As such, although programming at this level is conceptually simple (the programmer need only physically run through the task), it can also be extremely tedious, making inefficient use of both programmer and robot. Attempts at interaction with the robot environment tend to be difficult owing to the rigid structure of the recorded data, and similarly, programs are difficult to edit and usually too specific to be reusable. Naturally enough however, being the simplest programming systems to construct (as opposed to use), these **actuator-level** languages were historically the first to be used in industry.

Clearly, in the vast majority of cases, the relative positions of each joint of the robot arm is of only coincidental interest to the robot programmer. All that he usually cares about is the position of the end-effector with respect to the object it is working on. So long as the robot arm itself does not come into contact with any of its surroundings, then the actual configuration is normally irrelevant. Consequently, an **end-effector level** of programming abstraction has been developed which permits the programmer to describe movements not in terms of individual joint motions, but with respect to the required positions and motions of the robot end-effector. Such languages are also commonly referred to as **manipulator level** languages.

In such systems, the programmer specifies trajectory end-points and orientations, as well as specifying velocity parameters. Using this data, the controller then interpolates trajectories in real time. Although some systems allow the required **pose** (position and orientation) of the end effector to be specified mathematically, for instance using Cartesian coordinates, such information is usually conveyed by physically moving the end effector to the required pose and then using the internal-state sensors of the arm as a form of **digitiser** – in effect saying 'Here is the point I am referring to'. The controller usually then interprets the arm-joint positions as a corresponding pose for the end effector, and that is all that it then records. In some cases however, when high positional accuracy is specified, the controller may actually record each of the individual joint positions (as in actuator-level programs), and so avoid the possible errors caused by the mathematical transformations otherwise required. Except when this is done, the configuration of the robot arm when it later moves the end effector to a taught pose, may be different (and not easily predicted) from when the point was actually taught – it is only the position of the end-effector, not the arm, which is of concern.

End-effector-level languages include many programming constructs such as logical-branching primitives and subroutine facilities, allowing complex action descriptions and frequently even limited interfacing with simple sensors. As such, programming at this level is very similar to working on a conventional computer, and is finding ever increasing acceptance in industry. Because the required program complexities are frequently far greater than with actuator-level programs, such a similarity to nonrobotic computer languages allows the substantial experience of computing scientists in the field of software engineering to be of direct practical benefit. Thus, not only can programmers write general programs which are reusable, alterable and extendable, but large programs can be constructed which are still straightforward to follow and may even be written by a team of more than one person.

It is when a computational procedure becomes lengthy or complex, and therefore difficult to understand or describe, that it becomes particularly worthwhile to use a **structured** as opposed to **unstructured** language. Although both languages may be at the same 'level' of abstraction, the structured form has a sufficiently restrictive type of syntax to discipline the programmer in the organisation of his effort and division of the problem into simpler parts. In effect, by restricting the number of ways a program can be written, by controlling the logical 'flow' through the program, a structured language also restricts the number of ways a programmer can make mistakes!

The low and intermediate levels of abstraction already discussed are both termed **explicit** – that is, to varying degrees, they both require the programmer to specify explicitly what the robot must physically do. The highest level of abstraction, however, uses **implicit** programming techniques, where the robot system is made responsible for making some decisions on the sequence it must follow to accomplish a given goal, based on an understanding of the objects it works with. Programs at this level are only concerned with the relative positions and movements of the objects being manipulated during the task, not of the robot or even the end effector. Such **object level** languages (not to be confused with the computing term 'object code') employ symbolic **world model** representations of the manipulator, workspace, and objects involved in the task, to manipulate symbolically the objects to define positions, motions, and actions.

To support such modelling requires significant computing power, but proper utilisation allows powerful incorporation with CADCAM systems. In addition, not only can the robot itself be freed from any involvement in the teaching process, but programs can be designed to employ task-independent (and sometimes manipulator-independent) modules of subprograms, already stored in a 'library', which can then be 'custom-ised'. Nevertheless, because of the highly idealised form of the world models, interaction with the environment is not easy to program. Such object-level languages are still in the research stage, but it seems unlike-ly that they will replace end-effector-level systems completely for a very long time, and they may not even be used significantly in industry for several years. They are considered again in the final section of this chapter.

Although it is only second-generation robots which have external sensing devices, and work in this field is therefore largely experimental, nevertheless analogous levels of abstraction to mechanics software exist for programming of sensing devices. At the lowest level of abstraction, the 'sensor level', each individual sensor output is individually refer-enced by the programmer. At the intermediate 'parameter level' the programmer focuses on only a handful of meaningful features or para-meters obtained from **smart sensors**. At the highest 'object level', sym-bolic objects would be referenced through a world-model database, as with the mechanics software.

Textual programming

A tabular representation of the different methods of robot program-ming is shown in figure 6.7. Textual programming encompasses all three levels of abstraction covered in the preceding section, although textual actuator-level programming is really only found with certain simple teaching robots used for educational or hobbyist purposes. Such systems allow connection of the cheap robot arm to a general-purpose personal

implicit	object (world model) level	textual programming	structured/unstructured
	end-effector level		
explicit	actuator level	teaching by showing	walk through/lead through

Figure 6.7 The different forms of robotic programming

computer, and provide a simple syntax to permit the programmer to move a given joint a given number of degrees in a specified direction. Such languages all tend to be unstructured.

Textual languages used for industrial robots are nearly all of the end-effector-level type. Unfortunately, each manufacturer has tended to develop its own robot language, and this has resulted in little conformity and standardisation, and, of course, no universally accepted language. There are fundamentally two approaches which have tended to be used in the design of robot languages. The first method starts off with the details of the robot control system, and then constructs around this a syntax which will satisfactorily describe how the robot is to move. If necessary, additional language constructs are designed which permit such activities as logical branching in the control of the program. The advantage of a language designed in this fashion is that it is frequently possible to describe a manipulation task very neatly.

In many advanced robotic systems, however, a user program may ideally involve a great deal more than merely a collection of movement and sensing instructions. The software may contain sections for providing a suitable interface for an operator, system calibration procedures and calculations, error recovery routines and even limited data processing. In such cases, the requirements of the programmer include a substantial number of features commonly found in conventional computer languages, and a possible approach to language design is to start with an existing general-purpose language and extend it with the necessary constructs needed to control the robot system satisfactorily. Such an approach has the advantage that as the base language is already available, there is less initial design work to do, less chance of there being errors in the language, and less user documentation that has to be prepared from scratch (a not insignificant task).

Clearly, an important design feature of any textual language is the method by which users can interact with it. Languages tend to vary widely in how easy they are to learn and use, and how tolerant they are of user mistakes, yet making a language 'user friendly' is no easy task, and indeed is frequently a compromise between conflicting requirements. The nebulous term 'user' is commonly employed to cover anybody who interacts with the robot software, yet there are, in fact, a wide range of groupings within the term, all with different requirements and levels of skill. In increasing levels of sophistication, such groupings might be: machine operators, who tend to interact with the system by reacting to simple prompts, but may never actually see the robot program; maintenance personnel, who may employ special application programs which test the system, and who may have access to certain 'restricted' commands which must be 'hidden' from the general user; application-package users, who may specify the details of a particular task by using a simplified programming interface, possibly selecting from a 'menu' of options available on the screen; specific application programmers, who use a computer language in order to write the software to specify a particular robot task, probably employing several existing library subroutines; as well as application package writers, who provide efficient, robust subroutines for inclusion in libraries for use by other users.

Such a wide range of 'users' creates problems for 'user-friendly' systems – a prompt message which comes up on the terminal screen may seem a helpful reminder to one kind of user, but may become an irritating distraction to the experienced programmer. Similarly, the facilities a language offers may seem oversophisticated to the general user, but special language constructs for 'information hiding', error traps and so on may be vital for writing the very application packages that the general user then employs.

There are various textual languages which are currently used in industry. These include Unimation's VAL for their whole range of robots, IBM's AML for their asembly robots, Olivetti's 'SIGLA' for their Sigma robots, Automatix's 'RAIL', Bendix's 'TEACH' and Phillips' 'INDA', and there are also many equivalent languages that have been developed in research establishments, such as MIT's 'WAVE', which was among the pioneering work. Several of these primarily textual languages nevertheless support the use of a limited capability teach pendant.

An example of the continuing natural evolution of robot languages is VAL (Variable Assembly Language) which was originally unstructured, and bore a strong resemblance to BASIC. Recently however a new version of VAL has become available which is not only structured, but which is significantly enhanced, providing such facilities as external communication, sophisticated computational tools and concurrent execution of user programs.

As might have been expected from a computing giant such as IBM, AML (A Manufacturing Language) is a very sophisticated, well structured, semantically powerful robot language. Because IBM felt that no existing computer language had the right mix of features for robotic applications, they took the unusual step of first designing a completely new general-purpose language and then extending it with the suitable functions needed to run robots! This method they claimed allowed them to 'choose design trade-offs in a meaningful way'. The language itself is so arranged that subroutines which are found to be particularly useful can be 'upgraded' into new language commands recoded in the underlying sytem implementation without changing their functional behaviour.

As explained earlier, object-level languages are still very much in the research laboratories, and they include the University of Edinburgh's 'RAPT', Stanford's 'AL' and 'LAMA', and Purdue's PAL. The likely developments in these and other languages is covered in the next section.

Future software

It is debatable whether the current trend towards structured textual languages with large numbers of language constructs will eventually exclude all other forms of programming. Teaching-by-showing is a very easy method of transferring the manipulative skill of a human to a robot, and no end-effector-level textual language would ever be likely to compete with it. Clearly, an object-level language with sufficient understanding of a problem (such as paint spraying) could theoretically construct a sequence of movements for itself which was at least as good as anything a human could come up with. Indeed, the task might be one which no human alone had ever tackled, such as a complex coating process in outer space, such that an object-level language approach was the only feasible form of programming. Such an advanced level of object-level programming can, in fact, be considered as a separate **objective level** of programming (higher than the object level) in which artificial intelligence techniques are used to provide a degree of planning and problem solving ability. Nevertheless, it must be stressed that such sophistication is still some way off, although it is approaching fast.

In the meantime, practical programming is likely to remain split between teaching-by-showing and end-effector-level textual programming. Such textual programming may well not all become structured – it is undeniable that unstructured 'BASIC-like' progamming is often easier for the novice to learn. If the user is only ever going to interact with the system at a low level, then an unstructured language may be perfectly adequate for his needs, though it may get him into bad programming habits, causing problems if he ever needed to write complex software.

What might ideally develop is some form of standardised sophisticated language, capable of running on different makes of robot, and yet which was so constructed that a self-contained subset of the language could be employed by 'low-level users' without them needing to learn the intricacies of the whole language. In addition the whole software system should be designed in a modular fashion, with each level of software building on the capabilities of the lower level modules. Thus, at the lowest level might be the robot **operating system** while at the highest level would be user-generated software. Such an arrangement would provide inherent flexibility, so that substitution of any particular level (for instance a new operating system to allow the software to run on the latest available high-speed processor) would have no effect on the execution of the old programs of all the other levels. Similarly, replacement of one configuration of robot with another should not effect the execution of a taught program.

Since the development cost of software construction is increasingly becoming a major consideration when incorporating new robot designs, there are substantial advantages in the above system for both the robot vendor and the user. Consideration of programming costs are dealt with further in chapter 16. Nevertheless it is not at all clear just how such standardisation (which is universal in NC machines) will actually be brought about. It may well be that, rather than by international agreement, a language such as AML, through being pushed by a vast multinational concern, may in effect become a *de facto* standard for the whole robot industry, just as FORTRAN did for the computing industry decades earlier.

7

Sense and sensibility
EXTERNAL-STATE SENSORS

The principles of sensing

Within only a short time, robots will have gained many of the senses which it took humans millions of years to evolve. The sophistication of these 'robot senses' may not yet match their human counterparts, but rudimentary equivalents to human vision, touch and hearing are already common, and even a form of 'smell' is being used to sniff out leaks in car bodies. In this chapter we will be looking at **external-state sensors**, which to varying degrees furnish a robot with information about the conditions of its surrounding environment. It is the incorporation of such external sensors into a robotic system which qualify it for inclusion as a second-generation robot. First-generation robots include, of course, **internal-state sensors** such as encoders to measure the positions of the robot joints, as covered in chapter 5. Such **proprioception** (which in humans is a real 'sixth sense' which everybody has about body position) does not, however, provide any data on the robot's surroundings, and it is this distinction which separates the two robot generations.

Proprioceptive transducers are examples of **specific sensors**, which are included into the robot design for a specific task. Such sensors can be carefully 'tuned' for the situation they are to detect and can provide an unambiguous response with the minimum of computation. For external sensing, this approach may be chosen for certain tasks such as monitoring the speed of a conveyor belt, yet for many tasks a more flexible approach is employed using **general sensors**. Such sensing, for instance using vision, although more expensive, complex and difficult to program than specific sensing, can nevertheless be easily used for many different tasks, and because of its inherent flexibility can often be used to monitor and warn of malfunctions that a specific sensor would miss. Such an approach becomes even more vital in mobile robots, where the environment is highly unstructured and unpredictable.

Almost any kind of transducer can be interfaced to a computer-based robot controller – anything from a video camera to a simple microswitch. As explained in chapter 6, sensors such as cameras which produce a very large amount of data must usually be connected to a separate **preprocessor** which interprets the information and then informs the robot controller of the result. Such devices, which accommodate the stages of both sensing and early processing, and so output the information that is actually *required* from the sensing operation, are commonly referred to as **smart sensors**. Simpler sensors, however, can frequently be connected directly to the computer controlling the robot, and if the processor is fast enough, a large number of such sensors can in fact all be serviced by the one computer.

The ability of one computer to monitor several sensors arises from a technique known as **multiple input sampling**. Imagine an analog transducer which produces a signal of say between 1 and 2 Hz, possibly from measuring a vibration. Although the output voltage from the sensor is constantly changing, the computer monitoring the signal need not monitor it all the time. It can quickly 'sample' what the instantaneous voltage is at a given time, as shown in figure 7.1, then perform some other task (as with pseudoparallel processing), and quickly sample the voltage again before it has changed radically.

By interpolating between the different sampled output values, it is possible for the computer to reconstruct the complete signal as it would have been if it had been monitored continuously. In fact, theoretically,

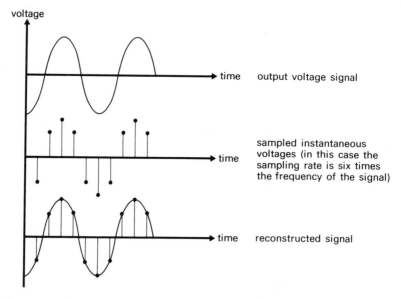

Figure 7.1 The principle of multiple input sampling

the signal need only be sampled at twice the maximum frequency that can be output from the transducer, so in the example above the output from the sensor need only be instantaneously looked at every quarter of a second. The **aperture time**, the time required to sample and store the amplitude of a signal, must be as short as possible if the value obtained is going to be accurate, and various techniques are available which can rapidly 'hold' an output value for subsequent processing. Clearly, using such **data capture** methods provides a lot of extra time during which the computer can control the robot or monitor other sensors.

Signals from all analog sensors are usually subject to **data shaping**. This may entail passing through a preamplifier (to boost the signal to the correct range), then through some form of **isolator** (which can be optical or even just a transformer for low frequencies), through various electronic filters (for instance to get rid of 'AC hum'), and then through an analog lineariser (to compensate for any nonlinear characteristics of the sensor being used), and finally through an analog-to-digital converter. In some cases this ADC may even be **multiplexed** between several different transducers. The **signal preconditioning** consisting of analog filtering and linearisation, may actually be replaced by **postconditioning** after the ADC using digital-circuit equivalents. As mentioned in earlier chapters, both the ADC and postconditioning circuits can even be replaced by a program written in software.

A common task is the monitoring of several different transducers to see if any has changed output, and this may be approached in one of two basic ways. With **polling** the controller sequentially tests each sensor to see if it has registered anything. If it has not, then the controller skips onto the next sensor; if it has, then the appropriate action is taken. Sensors which are more likely to change can be tested more often than others, but there is always the danger with polling that a sensor may change just after it has been polled, and consequently that the change will not be registered (however important it is) until the controller has polled every other sensor in the **skip-chain** (sometimes called a **daisy-chain**).

An alternative to polling is to use an **interrupt** mechanism, in which the processor does not monitor any sensor until an 'interrupt' signal is received indicating that one of the sensors has changed. The processor then goes into an **interrupt handling routine** which tests (possibly by polling) to see which sensor has caused the interrupt. Such a system is particularly suitable when there are only occasional signals from the sensors, as the computer is not tied up constantly monitoring them and can be used for other purposes until an interrupt is received.

By providing interrupt signals of different priorities, it is even possible for an important sensor, such as a collision detector, to interrupt all other sensors and direct the controller to service this interrupt in preference to all others, even if a different sensor has already started being serviced. With all multisensor systems however, it becomes necessary to 'weight' the responses from different sensors if they conflict. In a mobile robot, for instance, one sensor might suggest a large object was in the robot's path while another (different type of sensor) detected nothing at all!

Taction

Forms of sensing tend to fall into two broad categories: firstly, sensing by means of contact with an object, known as **taction** (or **contact sensing**); and secondly, remote sensing by means of such techniques as vision, collectively termed **noncontact sensing**. Taction can come in many forms, with varying degrees of sophistication, and currently, when employed at all, is nearly always used on the robot gripper.

At its simplest, taction entails **binary sensing** contact transducers, which can only register whether anything is touching them or not. Such devices may be straightforward microswitches, placed, for example, in the **fingers** of a gripper so that they are depressed when the gripper grasps an object. By configuring the switches into a small array, slightly more detailed touch information can be obtained. In place of switches some systems use a small number of force detectors and can therefore provide a graded, rather than binary response. Nevertheless, whether binary or continuously variable, systems with transducers at only a few points are often referred to as **simple touch** sensors, to distinguish them from full **tactile sensing** devices which consist of arrays of continuously variable force sensors which generate 'patterns' and are capable of providing a 'skin-like' response (of varying degree) which can distinguish physical surface features. The more advanced types of tactile sensing are still in the research stage.

Simple-touch sensors are already used in some practical robotic applications such as assembly, and contact probes are sometimes used for dimensional measurement. When simple binary switching is not used, stress and force are commonly measured indirectly by deformation of a resistance strain gauge. Such gauges consist of a thin strip of resistive material such as a copper–nickel alloy, etched in a pattern onto a thin flexible plastic sheet backing. Although the nominal resistance of the track on such gauges is about 100 Ω, when the plastic backing is bent or distorted the resistance of the track alters due to changes in the length of the track and the physical characteristics of the alloy. These small

Figure 7.2 Robotic assembly – potentially a major market for true tactile sensing

changes in resistance are then measured, usually by incorporating the gauge in a **Wheatstone bridge network**.

Such use of resistive materials in taction has been very widespread as the materials are cheap, heat resistant, and easily fabricated, yet problems with **hysteresis** and insufficient dynamic range suggest that future systems may use other technologies. Piezoelectric transduction, in which deformation of a crystal generates small voltages, suffers from the major drawback that it has no DC response characteristic, it only measures changes. Nevertheless, in some applications this may be acceptable, although at present the technology is rather too fragile for many industrial environments. Capacitance sensing is generally too susceptible to the influence of external fields to be practical, and in addition, like inductance sensing, its output is too dependent on the type of substance touching it.

True tactile sensing, although still largely experimental, is seen to have great potential in many areas such as factory assembly, fast adaptive grasping of arbitrarily oriented parts, handling of flexible materials, determination of weight and centre of inertia, as well as generally complementing vision systems. While vision may be used to locate and identify objects, tactile sensing is then seen as taking over for subsequent manipulation where force, pressure and compliance are important. A common example which highlights the importance of the sense of touch for a human is the task of screwing a nut onto a bolt. The nut is usually turned backwards until it 'clicks' and then carefully screwed forward, constantly checking that the threads have not crossed. Vision is not used at all. It is this level of sophistication that tactile systems aim to emulate by monitoring touch, force, pressure and slip.

The general approach to the design of such systems is to build an array of sensors on a thin, flexible substrate, like a robot 'skin' which can be placed over a gripper. The substrate should be compliant, and the entire structure very durable. It is thought that an array consisting of 10×10 force-sensing elements covering an area of, say, 5 cm^2 is sufficient for many tasks. Each element should have a rapid response time, with a sensitivity which allows it to register objects of from 0.001 kg to 1 kg, with low hysteresis. Many arrays have been built broadly along these lines, usually employing some form of conductive plastic mated with a grid of etched electrodes on a passive substrate. Contact forces produce local deformations which in turn cause the local resistivity of the sheet to alter predictably. This is then measured by the electrode grid and an 'image' is built up, to be processed elsewhere. In one example, however, developed at Carnegie-Mellon University, USA, the passive substrate has been replaced by a custom-designed **very large-scale integration** (VLSI) device that acts as a special-purpose parallel computer, so that the tactile sensor not only performs force transduction but also all the required 'image' processing as well!

Noncontact sensing

Proximity sensing

There is a group of sensors which, while not actually involving taction, nevertheless can only respond to objects which are in their very close proximity. The maximum range of this group of sensors is about 30 cm, but many actual designs can only operate over substantially shorter distances. Some of the technologies occasionally used for taction are also applicable to **proximity sensing**, such as inductive and capacitive systems. Inductive detectors can sense metals a few millimetres away, whereas capacitive systems can sense the presence of nonconducting materials such as wood, PVC or glass, as well as ferrous and non-ferrous metals over distances of up to a few centimetres. As mentioned in the last section however, both these types of sensor suffer from having very different responses to different materials.

A form of proximity detection occasionally employed is the use of jets of air which are deflected back on themselves if an object is close by. Such systems may even make use of fluidic logic systems to use this deflected jet to cause a required response directly. The most common designs of proximity detector, however, are based on optical methods, where either infrared or visible light produced by a transmitter in the sensor module is focused on a point in space several centimetres from the detector. A photosensor, located in the same module as the transmitter, is also focused on the same point in space. While there is no object at that point, the light reflected back at the sensor will be negligible. When a surface is near the point, however, substantial reflection will be obtained, with a maximum signal sensed when the surface is at the exact point of focus.

Such devices are only a few cubic centimetres in volume, and so can be physically located within the robot end-effector. They can also be very selective about distance measurement, but when this is not as important as increased range, the light is not focused to a point, and the detector merely registers the level of returned light, which will be dependent on the distance and the type of surface of the reflecting object. Such systems can register objects over tens of centimetres, but of course some objects will reflect light poorly; others will reflect well but because they are not placed 'head on' will reflect light away from the sensor! Similarly, transparent materials are a problem. Because, in addition to the transmitter light source, there may be other stronger lighting from elsewhere, all these sensors tend to **modulate** the outgoing beam (usually at a few kilohertz) and only register reflected light that is modulated in the same fashion, so filtering out extraneous signals.

Range finding

There are many occasions, especially with any form of mobile robot, when long-distance object detection is an advantage. Although proximity sensors and certain vision systems can provide this information, there are various approaches which are specific to range finding. Among the most popular of these is the use of ultrasonics. Such systems use the sonar principle of transmitting a brief burst of ultrasound (anything higher than about 20 kHz) from a piezoelectric crystal and then measuring the time it takes for any reflection of the pulse to be returned by a nearby object. Knowing the speed of sound in air (about 300 m/s) it is possible to calculate the length of the path travelled by the sound pulse, and halving this gives the distance to the object. After a massive research effort, Polaroid eventually developed an ultrasonic rangefinder for automatic focusing of their cameras. Because of economies of scale, and a special 'designers kit' which has been made available, the Polaroid sonar has become something of a standard for ultrasonic rangefinding.

However, in some other systems, higher frequencies of ultrasound allow greater resolutions to be obtained, because of the shorter wavelengths involved. Such systems may even have sufficient resolution to allow **depth mapping** of the surroundings in the form of contour maps. Lower frequencies, however, have their own particular advantages, because they are less attenuated by air and so can be employed to measure longer distances. The ranging accuracy of ultrasonic systems can be quite high, with an accuracy of the order of 1cm over a range of several metres using a 40 kHz signal, and using an audible signal of 1 kHz accuracies of 30 cm can be obtained over ranges of 50 m or more. The Polaroid system is supplied with a digital readout, and in practice is accurate to about 1cm over its range of about 0.3 m to almost 30 m. As supplied the sensor detects reflections within about 20° of its major axis, although this can be substantially reduced by setting it back within a horn. Alternatively, by using appropriate acoustic reflectors, it is possible to fan the beam out to nearly 90° from the major axis, and so obtain all round surveillance.

When multiple sonar systems are being used, some method of distinguishing between pulse sources is required. This can either be done by using different frequencies for the various systems (if the numbers are not high), or by using different pulse rates for the various transmitters. In these systems echoes are only accepted if they have the same pulse rate as that of the matching transmitter. Various refinements to simple sonar are possible, such as using phased arrays to produce a narrow-beam pulse in contrast to the conventional beams where the angular resolution tends to be more than 10°. Similarly, analysis of the complex waveforms produced by several objects which are not flat should be possible. However, such a task is unlikely to be any easier than analysing visual information! A simpler form of analysis may be that of sounds (including audible ones) from engines or possibly ceramics and glass, in an effort to recognize quality-specific patterns such as the 'ringing' of a piece of perfect glassware.

Despite their advantages, ultrasound devices can be prone to acoustic interference, and whereas some surfaces (especially sponge-like materials) tend to absorb the transmitted signal, any large flat surface acts as an acoustic mirror which, unless 'head on', can cause large surfaces to be missed entirely. Similarly, reflected signals drop off rapidly with distance, so that echoes from nearby objects may swamp those from objects further away. Finally, the factory cat may not entirely approve of all these high-volume pulses, which to it may be completely audible!

Other time-of-flight range finders using lasers or radio waves, as in conventional radar, tend to be unsuitable for the short distances involved in robotics, as the flight time is only a few nanoseconds per metre (although some space robots may find such systems ideal). However, microwave (or even ultrasound) **Doppler systems** may be used for safety purposes, as although they provide no useful range information they are very sensitive to relative movement.

Another technique for finding the distance of an object is to use triangulation, where in principle two devices are placed a known distance apart along a baseline, and the angle between the baseline and a given object is then carefully measured by each device. By comparison of the two readings, the actual distance of the object can be calculated. Optical sytems suitable for robots can either use **passive illumination** in which only existing ambient light is utilised and triangulation is by **stereopsis** (covered in the next section), or **active illumination** techniques, which avoid the difficult problem of image registration.

Such an active-illumination approach uses a slowly scanning collimated light source (such as a solid-state infrared emitter or laser) to cause a bright spot of light to move over the surface of nearby objects. This spot is then detected by means of a rapidly scanning sensor (the light source is usually modulated to aid discrimination between it and background lighting) so that a series of triangulation readings are built up as the spot scans across the surroundings.

Vision

Just as sight is the primary sense used by humans in many applications, so too robot vision is destined to become a vital sensory input in advanced systems. The machine vision industry is currently estimated to be doubling in value every year, becoming worth $1 billion per annum by the end of the decade! Vision offers many advantages over conventional senses, being potentially fast yet leaving the workpiece undisturbed. Being a truly general sensor it can provide a range of information about the scene viewed, from recognition of the type, location, range and orientation of objects, to complex feedback information for guiding a manipulator or vehicle during a given task. It can even be used for visual inspection and quality control.

The techniques used in robot vision systems can be very involved, and it is only possible to provide a broad overview in this section. It is also worth remembering that much of the image processing is also applicable to other sensors such as tactile arrays. A lot of the early robot vision work came straight from standard computer vision research, which itself was stimulated as a result of NASA's interest in vision for space exploration. However, there are two important differences between the two fields of computer and robot vision. Firstly, with robot vision the field of view may be constantly changing, and secondly, in order to be useful the vision system must work in nearly real time.

The earliest robot vision system was probably 'SIRCH' (Semi-Intelligent Robot for Component Handling), built at Nottingham University, UK, in 1973. It consisted of a movable gripper under control of a computer which interpreted positions of simple shapes placed in the field of view of a camera. Objects could be detected and manipulated within about five seconds. Unfortunately, lack of commercial interest in the prototype resulted in the system never being developed into a marketable product.

Cameras used in robot vision can either be placed above or to the side of the robot, or can actually be attached to the robot wrist itself, in an attempt to obtain a clear view of the work area. Developments in **fibre optics** also allow cameras to be situated a small distance away from the wrist and the image transmitted by means of a **light guide**. This even enables images to be obtained from locations where no camera could be placed, such as in the 'palm' of a gripper. The cameras themselves are either ordinary TV camera tubes (such as vidicon cameras) or else solid-state cameras (such as **charge coupled device** or CCD cameras) which consist of a mosaic (or sometimes just a line, as in figure 7.3) of individual photosensitive elements.

Although CCD cameras are a relatively new technology, it is expected that because of their superior reliability and their size, weight and power consumption characteristics, they will soon largely replace conventional camera tubes for robot vision. It is quite possible to use simple CCD's consisting of a single row of photoelements to build up a complete two-dimensional image, if the **linescan camera** is placed across a conveyor belt. As objects pass beneath the camera, it builds up an image in its memory slice by slice.

The resolution of a vision system is determined by the number of individual dots (or picture elements, **pixels**) which can be distinguished per image, and it is usually no higher than 512×512. Images with resolutions below 64×64 pixels tend to be severely degraded by the digitisation process. The distortion present in the image (the **geometric fidelity**) can be quite large with conventional TV tubes, but with CCD's, because of their inbuilt structure, the fidelity is very high.

Whatever camera is used, the analog signal from it must be digitised using an ADC, and the information stored so that it can be processed by computer. Storage is commonly a **frame buffer**, which is simply some

Figure 7.3 Typical example of a solid-state camera

fast computer memory for which each memory address corresponds to a pixel. Each pixel in an image may be a simple binary element of black or white. On advanced systems, however, each pixel may be digitised to any one of 256 **grey levels**, ranging from pure black to pure white. In the future, robot vision may employ colour information instead of simply levels of brightness. At present, however, it is debatable whether in most applications the small additional advantages of using colour justify the necessary increase in computer processing.

To save processing time, only a few robot vision systems use grey-level information, and instead they preprocess the image to produce a binary version. By selecting a threshold level of brightness, each pixel is made black if below the threshold and white if above it, so producing a silhouette. Clearly, lighting of such systems is all important, as shadows on an object might be interpreted as background, so radically altering the object's perceived shape, and highly reflective objects might also cause spurious effects. Changing the threshold can also dramatically affect the system's interpretation of an object.

To try to overcome such problems, objects may be lit from beneath, or if that is impossible tubular fluorescent lamps may be used to provide uniform illumination. In some cases it may be feasible to place objects onto backgrounds coated with fluorescent materials, and then illuminate them with ultraviolet light to create a high contrast between the background and any nonfluorescent object lying on it. Coloured filters in front of the camera lens may further enhance contrast. Such care over passive illumination is vital if clean images of features of interest are to be obtained, and the overall reliability and efficiency of a given vision system is frequently determined by how successfully this is done.

In humans as well as robots, there are various methods by which objects in complex images can be distinguished from their backgrounds. The edges of the objects may be revealed by changes in colour or brightness, or they may be inferred because of the background that they 'blot out'. If account is taken of how the image changes over time, then objects can be distinguished because of relative movement between them and their backgrounds, either because the two are in fact moving relative to each other, or because the observer is moving, causing an apparent change in position of the object due to parallax. Instead of the observer physically moving to obtain two views of the same object, the same parallax effect can be obtained by using stereovision (stereopsis) as a human does (or in some cases even more than two views can be used).

Of these methods, the most common form of **low-level feature extraction** used in sophisticated robot systems using grey-levels involves **edge**

detection. The first stage of such image processing systems is **segmentation** in which images are split up into regions with common brightness or colour. Edge detection itself involves measuring all the changes in brightness across an image. Where there is a rapid change in intensity, this may correspond to a physical edge and it is noted in an **edge map**. Special hardware is available which can carry out all of the edge detection process at high speed. An alternative approach which is sometimes used is **region growing** in which an attempt is made to isolate regions in the image corresponding to surfaces in the original scene, by partitioning together regions of a similar brightness.

Because the supposed 'edges' in an edge map are typically short discontinuous line segments, **edge clustering** is performed to join them into continuous lines which should correspond closer to the actual edges of the object. The results of this process are then stored in a more compressed form of data representation such as **chain coding**, in which the lines are described by giving, for each pixel on the line, the direction (one of only eight possible directions) of the next pixel.

The final, but most difficult, stage of any robot image processing (whether of binary or full grey-level images) is **pattern recognition** (or **scene analysis**), in which the image must finally be interpreted in terms of objects and their relationships to one another. Somehow, features of the image must be matched to features of the objects the system 'knows about'. Two common approaches are: **template matching**, in which the chain code of the image is matched against a set of 'templates' of possible objects and the best match is selected, and **feature discrimination** in which a set of derived higher level features such as area, perimeter, and centre of area are matched against a set of corresponding features of known objects. A great deal of effort is being expanded to try to develop AI techniques which can improve such approaches. At Brunel University, UK, however, a highly sophisticated vision system called 'WISARD' is already sufficiently advanced to be able to distinguish whether the face of one of its inventors looks happy or sad – a voice synthesiser then says, 'Hello' in an appropriate tone!

Rarely in practice can problems such as overlapping components be easily avoided. The task of picking a single part from a jumbled mass in a bin (the so called **bin-picking problem**) to many became symbolic of the level of sophistication which robotic vision should aim at. The problem was first solved at the University of Rhode Island (which has a long history of advanced research into robot vision), but the technique still requires substantial improvement before it will be suitable for industry.

Depth information of a 3D image can be obtained visually using either active or passive illumination. When no special lighting technique is used to help in ranging, differing viewpoints of the scene must be obtained, either by using two (or more) slightly separated cameras, or by using one camera which can be moved. In either case, corresponding points in each image must somehow be matched (a far from simple task) and, knowing the angle to the camera each point represents, their distances can then be calculated by simple triangulation. Some image systems which are fundamentally 2D, but make use of a limited amount

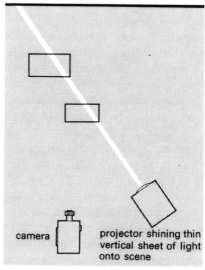

Plan of System

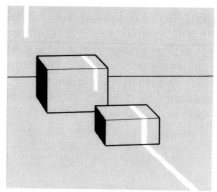

Camera's View of Scene

Figure 7.4 The principle of structured light

of depth information in order to distinguish objects, are commonly referred to as **2½D** systems.

Active illumination triangulation systems, similar to those covered in the last section, have a camera at one end of a baseline and a scanning laser (or equivalent) at the other. The camera can easily detect the bright spot of the laser, and determines the effective angle to it by the pixel it falls on. Correlation of this angle with the current angle of the scanning beam enables the distance of the spot to be triangulated. In some such depth-mapping systems, instead of the laser constantly scanning the scene, it is steerable, and can be directed towards 'points of interest' to determine their range.

Another similar form of active illumination used in some robot vision systems is **structured light**. Such systems project a thin sheet of light (instead of a single spot) on to a scene (figure 7.4), and so in effect obtain a large number of triangulations at one time. The same approach has been used to obtain clear binary images of thick objects on a conveyor belt passing beneath a linescan camera. A sheet of light is shone in a plane which is at an angle to the plane of view of the camera, but it is arranged that the two planes intersect on the conveyor belt in a line exactly beneath the camera (figure 7.5a). Where the conveyor is empty, the camera sees a bright line of light, but where there is a thick object underneath, the line appears, from above, to be deflected out of the view of the camera (figure 7.5b). This principle is employed in the 'Consight' system, developed by General Motors in 1977, which uses *two* sheets of light projected from different locations. It may well be that elegant use of simple techniques such as these will largely have to serve

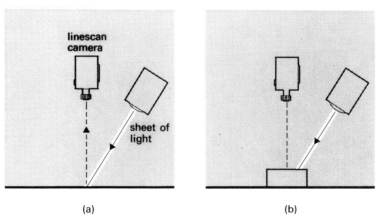

(a)   (b)

Figure 7.5 The principle of the 'Consight' system

in industry, until cheap and fast yet sophisticated vision truly becomes available.

At the moment, several manufacturers are selling separate robot vision systems, but these tend to cost around £25,000 – maybe half or more of the cost of the actual 'blind' robot to which the system will be attached. ASEA, however, have recently introduced what is claimed to be the first 'integrated vision system', working under actual factory conditions, which can be simply programmed by factory shop-floor workers. It has been estimated that in 1981 only 1% of USA robots employed vision, but that by 1990 the figure will be closer to 25%. The days of the 'blind' robot are already numbered!

8

Hand in hand

END-EFFECTORS AND
PARTS PRESENTATION

END-EFFECTORS

Types of grippers

A robot can be thought of as nothing more than a convenient means of moving a device attached at the end of the robot arm to different locations. Such 'end effectors' which can be placed on the end of a robot come in may different forms, but can basically be categorised into grippers, tools and sensors.

The majority of end effectors are different types of gripper devices designed to allow the robot to pick up objects. Frequently grippers can not only grasp workpieces but also centre and orientate them, and more advanced devices include some form of sensing, at least to indicate if a part is present or not. On the whole a gripper should be as light as possible, for the maximum payload of a robot includes the weight of an end effector, so the heavier the gripper the less that is left for any workpiece. Similarly, the smaller the gripper the 'tighter' the work-spaces it can squeeze into. If an object is to be held accurately then the gripper must be as rigid as possible and must grip with sufficient force to prevent slipping (though too high a force may cause damage to the workpiece). Heavy loads should be held as close to the axis of grip as possible, so as to avoid high **moments** on the gripper and robot. Finally, to be cost-effective. the gripper must be easily designed, manufactured and maintained, and must be reliable.

Flexibility of use for robot grippers is currently very low, and it is usual for grippers to be specially chosen or even designed for a particular industrial application. This results in costs which are typically up to 10% of the robot price – a substantial 'hidden cost' of buying a robot which many users have been caught out by. The most common form of gripper involves some sort of clamping mechanism similar to a pair of pliers, as shown in figure 8.1, in which easily exchangeable **fingers** grip

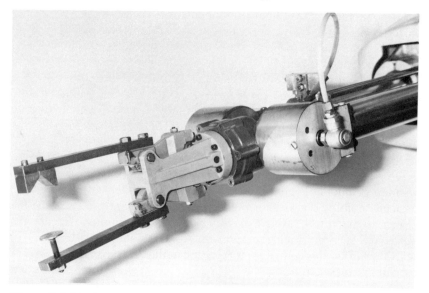

Figure 8.1 Typical form of general purpose robot gripper

components either externally, or by 'opening up' internally. Of such grippers, the two-fingered varieties dominate, using pneumatic or hydraulic actuators to operate them. Although hydraulics allow a high clamping force, it is more difficult to adjust the level of the force than with pneumatics. Some pneumatically actuated grippers employ a mechanical toggle-type mechanism to clamp parts, multiplying the force from the piston by up to twenty times. Electrically operated **proportional grippers**, capable of closing to a specified position (rather than being either open or closed) are slowly becoming available.

To prevent a part slipping or pivoting in the gripper jaws, fingers tend either to be covered with resilient pads or else to use some form of self-alignment to ensure that each jaw has at least two-point contact with the object being handled. Care must be taken in choosing a gripper design so that there is sufficient **stiction** to hold a part firmly even during acceleration of the robot arm. When a part is lifted at full speed, the force acting on the part may be two or three times its normal weight. The most commonly used material for gripper pads is polyurethane bonded to steel. The polyurethane has a high coefficient of friction and can be conveniently machined (after being frozen) to an appropriate shape. It will also stand up to thousands of hours of compression and release. Clearly, however, the choice of suitable materials for fingers can be problematic when hot components must be handled (as in forging).

Grippers for large yet light components, such as cardboard boxes and cartons, may use two independently controlled jaws placed a wide distance apart, whereas smaller, heavier, flat or box shaped objects may require a special form of linkage mechanism which causes the jaws to move together while their surfaces remain parallel to each other. Fragile components such as sand cores or glass bottles may be gripped not by moving jaws but by some form of expanding bladder or balloon placed inside or around the object and then inflated. The resulting high contact area allows very low contact forces for a given weight, and will not scratch or mark a surface.

Objects with surfaces that are even, dry and clean may be suitable for **vacuum grippers**. There are many variations to such grippers, but all basically employ one or more suction cups to attach to a surface of the object to be lifted, using a **venturi** or vacuum pump to create the necessary suction. Light components or components with sticky or polished surfaces may require **forced unloading**. This is usually achieved by using a **blow-off system** which reverses the direction of the suction mechanism, so building up pressure in the cups to force them away from the object surface. The same approach can be used with any type of surface whenever rapid release is required. A major advantage of vacuum systems is that they can be used when only one surface is available for 'gripping'.

Ferrous objects can be lifted using magnetism, usually from an electromagnet, with configurations and conditions similar to those used with vacuum grippers. Care must be taken, as there is always a tendency for flat surfaces to slide at right angles to the magnet, and metal swarf and the like may be attracted just as much as the desired object! Permanent magnets may have to be used in potentially explosive environments or at temperatures above about 60°C, but they require some means of either forced unloading or else displacement of the magnetic poles (as used, for instance, in special 'magnetic tables'), which may induce permanent magnetism in the handled parts. Electromagnets avoid such problems by demagnetising with a simple short reverse of polarity or by using a burst of alternating current.

Additional methods which are occasionally used for picking up objects include: **adhesive grippers**, in which sticky fingers pick up light materials such as cloth, possibly using a continuously advancing ribbon of sticky material (similar to a typewriter ribbon); **puncturing grippers**, in which the gripper pierces the object or material to be lifted; use of scoops, ladles or spatulas; as well as simply hooking onto a part. Examples of some of the uses made of the grippers covered in this section are contained in chapter 9.

Flexibility

Gripper design for the handling of a given workpiece may be difficult enough, but as the number of different parts to be handled increases so too do the often conflicting design requirements. Although fingers can be designed to hold different sizes of object at different points along their length, in many cases such an approach is not feasible. As a result, some robots use a **double gripper** consisting of two independent grippers on the one robot wrist (such as that shown later in figure 9.2). By rotating the wrist, either gripper can be brought into position, so that, for instance, one gripper might be used to place a part into a machine tool but the other gripper used to subsequently remove the (differently shaped) part.

Such systems are satisfactory when only two different shapes are to be handled, but when several variously shaped objects are present, a **change-over system** is often employed. These systems automatically replace either part or all of the gripper for another. This may be accomplished by attaching different grippers to a 'stub' on the robot wrist, or else one gripper may remain permanently attached and be used to 'pick up' the alternative grippers. Although expensive to design, build, maintain and program, change-over systems may be suitable if the robot can accomplish the swap during a period in which it would otherwise be idle. Unfortunately however, in tasks such as assembly the robot tends to 'pace' the whole system it is working with, and any time spent other than actually assembling is time wasted which is immediately reflected in an increased final cost of assembly.

In such circumstances, some form of **universal gripper** is required which can pick up a wide variety of different parts. Various such **form-adaptable grippers** have been developed – for example at MIT (USA), the Cranfield Institute of Technology (UK), and IPA (Stuttgart). A common approach is to employ three or more (often independently controlled) jointed fingers, analogous to a human hand. It is tempting to try to mimic the human hand directly, but as this requires 22 **degrees of freedom** it is a somewhat formidable task! In practice, it seems that only three fingers can in fact handle almost as many parts as five, whereas two fingers are less than half as effective. Consequently, most designs employ just three fingers, with varying degrees of sensory feedback.

A passive approach to form adapting uses a conventional gripper that has fingers padded with a loose lining filled with granular powder. When an object is gripped, the padding moulds around the object surface and is then 'fixed', prior to moving the gripper by locking all the granules

Figure 8.2 A form of universal gripper – the 'Omnigripper', built at Imperial College, London

together. This may be accomplished either by electromagnetising granules made of iron, or by applying a vacuum to the filling. However,
with both this passive approach and with the **anthropomorphic hand**, it
is difficult to sense, let alone predict, where exactly a grasped object is
relative to the robot wrist.

A new design, built at Imperial College, London, which seems to
overcome some of these problems is the 'Omnigripper' shown in figure
8.2. This gripper has two parallel slightly separated 'fingers', each consisting of an array of 8 × 16 closely spaced pins which can ride vertically
up and down independently of each other, as in figure 8.3a. Lowering
the gripper over an object pushes some pins up out of the way, so
creating 'customised' fingers to fit the part. To grasp objects, either the
two slightly separated fingers can be brought together, as in figure 8.3b,
to grip an object externally, or they can be moved slightly apart (from a
'closed' position) for an internal grip, as in figure 8.3c.

Feedback from each pin of the Omnigripper provides taction information about an object (including height details), and this allows pattern
recognition of objects, compensation for inaccurate part positioning and
sensing of external surroundings (because those pins not actually hold-

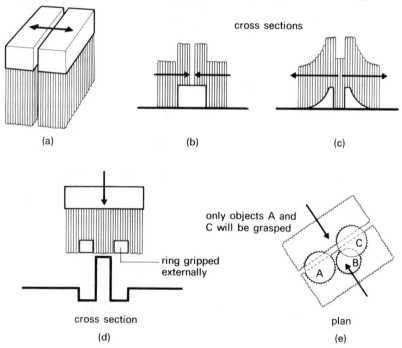

Figure 8.3 Principle of operation of the Omnigripper

ing the object will still be pushed up if they touch anything new, as in figure 8.3d). In addition, the Omnigripper can pick up more than one object simultaneously, and can even be used to select and separate overlapping parts, because only those parts surrounded by pins from both fingers will in fact be grasped, as in figure 8.3e; others will merely be slightly moved.

Compliance

In several parts-handling robot tasks it is useful to build some 'give' (or **compliance**) into the gripper which will accommodate slight variances in the positioning of parts. Such positioning errors can all be classified as combinations of either 'linear' or 'angular' errors, as illustrated in figure 8.4, which shows the simple task of placing a peg shaped object into a hole. In this example (which could just as easily be placing a part into a machine tool) large lateral errors can result in the peg missing the hole, while, once the peg has started to enter the hole, angular errors can

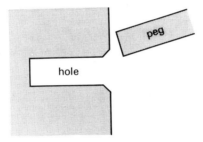

is a combination of:

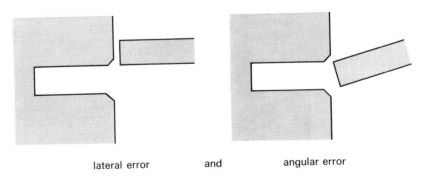

lateral error   and   angular error

Figure 8.4 Any positioning errors are a combination of lateral and angular errors

cause either jamming or at least far higher contact forces than would otherwise occur.

Provided that the peg is gripped sufficiently compliantly, then small lateral misalignments can be corrected by allowing a chamfer to 'guide' the peg to the hole. Some robot systems use **active compliance**, where sensory feedback (nearly always of force) is used to reduce lateral and angular errors by physically driving the gripper to produce compensatory motions. The force sensors for such active compliance are usually located in the gripper itself, although some robot designs incorporate them into the joints of the robot, however, the speed with which all systems can react is dependent on how high a contact force is permissible.

In the search for faster, but simpler designs, much work has been done on **passive compliance**, using purely mechanical solutions to the problem. A common misconception is to believe that it is always sufficient merely to use the inherent compliance of a 'floppy' robot arm (given how hard it is to build a stiff one). This may sometimes be the case, but such forms of **structural compliance** are not always suitable. Imagine a peg which is angularly aligned with a hole, but slightly off centre. If the peg is simply held 'loosely', as above, it will hit the chamfer as it descends, developing a moment which knocks the peg off angular alignment. As it enters the hole this will cause **jamming** (which is locking retrievable by changing the direction of the applied force and moment) or **wedging** (which is irretrievable locking requiring that the peg be pulled out and reinserted).

The form of compliance which is in fact required would correct the lateral error without introducing any angular error. The comparatively recent SCARA-**type robots** can provide such characteristics, because they exhibit large lateral structural compliance yet small angular compliance along the horizontal direction, and small lateral (although high angular) compliance, in the vertical direction; so making them suitable for insertion tasks in assembly. Of course, such designs cannot cope with horizontal angular errors as well as lateral ones. For these cases, there is a radically different approach, involving purely mechanical multiaxis 'floats' known as **remote centre compliance** (RCC) devices, shown schematically in Figure 8.5, which can be attached to the end of any robot wrist as a platform for the gripper itself.

RCC's are designed so that any angular error sets up a moment which causes one part of the device to rotate about a 'remote centre' (usually located at the end furthest from the gripper of the object being held), yet does not cause any lateral movement. A separate section of the RCC

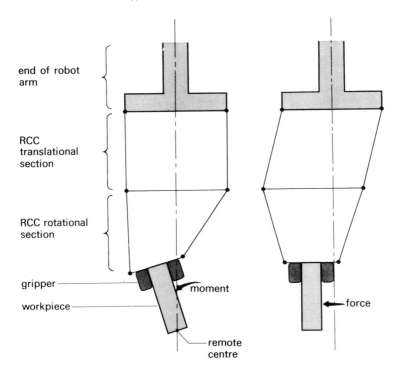

Figure 8.5 Principle of operation of a remote centre compliance device

allows the first section to move laterally independently, without any rotation of the workpiece. Such an arrangement results in positioning characteristics which are the same as if the peg were being pulled rather than pushed into the hole.

Practical designs of RCC come in many forms, ranging from devices using sprung levers to those using pneumatic tubing to provide the correct response (such as that shown in figure 8.6). However, all the designs allow the use of minimal force to achieve close tolerance fits (particularly useful in assembly work), permit greater variability of parts tolerances and cut down machine stoppages due to jamming and wedging. Far higher tolerance insertions can be accomplished than a given robot could achieve alone, for lateral errors can be as large as the chamfer used, and even press-fit operations can be managed in which the (deformable) peg used is actually slightly larger than the hole into which it is forced.

Figure 8.6 Example of a remote centre compliance device without end-effector attached

Tools and sensors

In addition to attaching various forms of gripping device to the robot wrist, it is common for various types of tools, and even just sensors, to be used as end effectors. As with grippers, it is quite possible to arrange for different tools or sensors to be automatically changed by the robot itself. Typical tools include spot-welding guns, arc-welding torches, cutting torches, heating torches, spray guns, adhesive and sealant dispensers, **nut runners**, drills and countersinkers, engravers, impact wrenches, grinders, belt and disc sanders, water jets and lasers. Because of the high power of the lasers commonly used, they are frequently larger than the robot using them. As a result, special **light guides** are built which direct the laser beam along each articulation of the robot arm, and out through a projector at the wrist. Further details of this, together with examples of the uses of the other tools mentioned above, are covered in chapter 10.

Although it is not yet often economic to dedicate a robot purely to sensing data, it is occasionally (and increasingly) useful to use an end-effector which is solely a sensor. A well publicised example is the use of a 'helium sniffer' at Austin Rover's Cowley plant to detect leaks from a sealed car during production as part of a waterproofing process. Sensory probes are occasionally employed to determine the accuracy of physical dimensions of products, and ultrasonic contact sensors can be used to detect flaws such as hairline cracks. It is to be expected that, as more and more of the industrial process becomes automated, so the role of the 'inspection robot' which monitors that automation and preserves quality control, will likewise steadily increase in importance.

PARTS PRESENTATION

Feeding workpieces

As robots are increasingly employed in sophisticated systems, whether for machining parts or for subsequent assembly, so too the need increases to understand the various alternative methods of presenting components to a robot. Many of these techniques have in reality been directly inherited from the more traditional approaches of production engineering, and it is debatable just how suitable some of the techniques are for robotic work. Nevertheless, not only are these the methods with which industry is most familiar, but, until new alternatives are devised and proven under industrial conditions, they are all that the roboticist has at his disposal.

The range of devices must somehow cover all the different shapes of components which may be used. To accommodate this requirement involves having many different approaches available, and then selecting the most appropriate method for the particular component involved. Although a wide range of workpiece shapes may be encountered in a robotic system (especially in assembly), it is possible to group such shapes into 12 broad categories, based on similarity of suitable feeding techniques (figure 8.7). Parts may be: interlocking, such as helical springs; primarily flat; cylindrical; prismatic; fundamentally conical; pyramidal; headed (including all simply stepped versions of the last four shapes), such as screws; hollow, such as box sections; complex rotational, such as crankshafts or camshafts; irregular solid, such as some pressings and forgings; long, like steel strip; as well as approximately spherical.

Although there are a large number of potential methods for handling such a wide range of part types, in practice there are relatively few principles that are actually frequently employed. It is these approaches which are dealt with in the following sections.

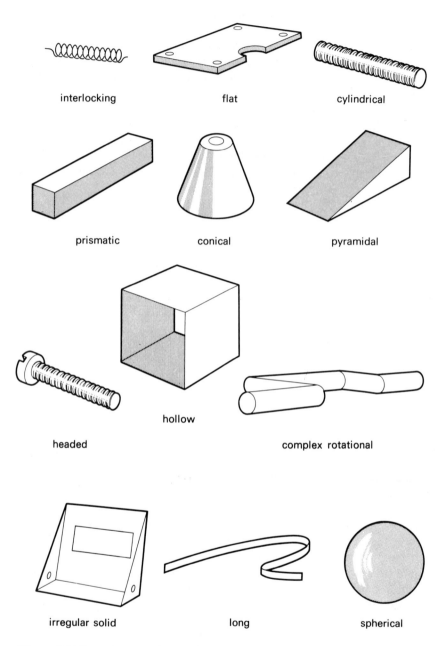

interlocking   flat   cylindrical

prismatic   conical   pyramidal

hollow

headed   complex rotational

irregular solid   long   spherical

Figure 8.7 Broad categories of different component shapes

Transfer systems

Whenever objects must move between separate workstations, whether manual, automatic or robotic, some form of transfer mechanism is required to move the base (or **work carrier**) on which each object rests. Such a system must ensure that no relative motion occurs between the carrier and the workstation during an operation, and that the correct object orientation is maintained between stations. Recently, there has been increasing interest in the use of **automated guided vehicles** (AGV) for this task (discussed further in chapter 12). However, in the majority of cases, each work carrier does not have its own independent transfer mechanism, but is instead part of a larger transfer system.

Such systems are commonly classified according to their layout and continuity of operation. **Continuous transfer** systems move the work carriers past the workstations at a constant speed. An automatic workhead or a robot must somehow move 'with' the object it is currently working on until the operation is finished, and then rapidly return to its initial position to start work on the next object. With **intermittent transfer** however, the workheads remain stationary and it is the work carriers which start and stop in front of the workheads to allow time for the required operation.

If all the work carriers on a system move at the same time, then the system is said to be an **indexing** machine. With such an arrangement, any unscheduled stoppage of an individual workhead forces the whole system to stop. With **free-transfer** machines, on the other hand, workstations are separated by buffer stocks of workpieces, and transfer to and from a given buffer occurs only when the required operation on an object has been carried out by a particular workhead. With this arrangement, malfunctions at one workhead will not necessarily hold up the whole system, as it is likely that limited supplies of workpieces will still be available for the other workheads from adjacent buffers.

All transfer systems can be built in a straight line, termed **in-line** systems, but some provision must be made in these cases for the return of empty work carriers from the end of the line (where finished products are removed) to the beginning. An alternative to this layout, which is applicable to continuous transfer machines and indexing machines, is a **rotary** configuration, in which a circular 'table' rotates to bring one work carrier after another under the workheads which are stationed around the table's circumference. A given workpiece is completed after having passed under each workhead, so that each time the table moves round by one workstation another finished product can be removed from the system.

Typical conveyor mechanisms for in-line systems include driven roller conveyors, 'walking beam' conveyors (in which a lifting beam regularly lifts and moves objects slightly forward on support beams), chain-driven work carriers (where all the carriers are linked onto a driving chain), and shunting systems (in which all the work carriers are touching end to end and a mechanism pushes the end of the line of carriers).

Parts feeders and storage

Components for tasks such as assembly tend to be stored in an un-ordered state in some form of bin or hopper. Such containers include some kind of unloading system, but clearly an additional stage is required to orientate the parts. An alternative approach to parts storage which does not require subsequent orientation is the use of magazines. With this method the components are packed in an oriented fashion into a special container, and are delivered with the aid of spring loading, compressed air or simply gravity. A special case of the magazine is the **pallet**, in which components are stored in a horizontal arrangement on a supporting board, usually in a series of 'nests'.

Magazines offer various advantages over other forms of parts feeding, as they can be very efficient, can often dispense with any need for feed track and frequently reduce downtime due to blockages. On the other hand, considerably fewer parts can usually be held, and magazines can be very expensive to manufacture. The ideal place to fill them is at the site of component manufacture, but this may cause problems in trans-portation. In some cases it may be possible to employ disposable maga-zines (similar to those containers used for certain medicinal tablets in which each tablet is held separately). Alternatively, parts may be linked together in bands (as electrical components are for circuit-board assem-bly) or, for parts which are punched from a strip, it is possible to leave slight links between components which are only finally broken when the part is required. Research into such technology is being conducted at the Cranfield Institute of Technology, UK. For small parts, the whole punching operation may be left until the part is required, although there may be problems in matching the rates of production and demand.

There are many alternatives to using magazines, and of these feeding devices the most versatile is the vibratory bowl feeder (figure 8.8). Although its technology, which was developed for fixed-automation applications, may be something of an 'overkill' for most robotics work (it tends to be capable of delivering parts, at a cost, far faster than a robot needs), nevertheless it remains a very common solution. These

Figure 8.8 A complex vibratory bowl feeder designed to feed matches head first on 13 parallel tracks

feeders consist of a helical track wrapped around the inside of a shallow bowl. The bowl is vibrated by an electromagnet in such a way that, when components are placed inside the bowl, they have a tendency to slowly climb up the track to the outlet at the top.

One of the great advantages of these feeders is the relative ease with which orienting devices can be incorporated within the bowl. Such **in-bowl tooling** usually consists of specially designed tracks on which the components pass various constrictions and cutouts which allow through only those parts which, by chance, are being fed in a correctly oriented fashion. In **passive orienting devices**, parts which do not have the desired orientation are rejected by letting them fall back into the bowl, so recirculating them to make another attempt to pass through the orienting system. **Active orienting devices** on the other hand, include special channelling designed to reorient the parts. Such devices tend to have higher output feed rates than the equivalent passive system.

A major drawback of traditional feeding systems is that they exhibit far less flexibility than the robots they are used to feed. Although bowl feeders can be tooled to orient many different parts, the design and manufacture of such tooling is a skilled and expensive task. In addition, as mentioned earlier, the overall cost of such feeding systems may not be justified by the comparatively low feed rate that the robot actually requires (as opposed to what the feeder can in fact deliver). Attempts have been made, for instance, at Salford University, UK, to devise simpler feeding devices which, being slower but cheaper, would more ideally suit robot applications. By arranging for the special orienting tooling to be easily replacable, it is even hoped that the same CADCAM system which generates a particular component could also automatically produce the necessary channels needed to subsequently orient it.

An alternative to this approach is to use general purpose bowl feeders, without any special tooling for orientation, but with some form of pattern recognition system which can recognise the shape, position and orientation of a part and then either reject incorrectly oriented components or direct some form of active orienting system to align the part. Such systems, being inherently flexible though expensive, can be used to load different magazines with a variety of parts rapidly, so that only one feeder is needed to service the requirements of a robot system which otherwise would have required as many different feeders as there were components used. A variation on this approach (which has been incorporated by Westinghouse into the APAS system mentioned in chapter 17) is to employ special **programmable feeders** in which the tooling can be physically adjusted automatically, under computer control, in order to accommodate changes in part styles.

Part III
ROBOTICS IN ACTION

By definition robots are flexible. This results in a wonderfully wide diversity of potential applications, the range of which is increasingly restricted only by the imaginations of those empoloyed to think of them. The four chapters in this section explain not only the major existing industrial uses for robots, but also detail the problems involved with robotic assembly, which is expected to become the single largest robot application within a decade. In addition, consideration is given to some of the applications for nonindustrial robots, and for robotic-like equipment which in some countries is, in fact, thought of as being robotic.

9

Robots with hands

WORKPIECE MANIPULATION

This first chapter of existing robotic applications deals with those that are accomplished using some form of gripping device attached to the end of the robot arm. Of necessity, those workpiece-handling tasks which have been successfully robotised are those which on the whole have been ideally suited to first-generation robots, without any significant sensory feedback from the environment. In addition, because the robots in these applications have tended merely to be employed to move objects from one given position to another, with no specified requirements for the actual path followed, most of the robots have only needed to employ point-to-point controllers.

It seems likely that, even when more advanced robots are readily available, for at least some of the tasks dealt with in this chapter, industry will continue to employ only first-generation robots – often the only level of sophistication required. New robots will bring with them new applications and solutions to only partially solved older applications, but it will rarely be economic to use them to supplant older models in instances where the outmoded technology is otherwise perfectly satisfactory. (The broad area of robotisation in industry is dealt with in Part IV of this book.)

Robotic assembly is an example of an application which was infeasible before second-generation robots became available. Despite being a workpiece-handling activity, however, it has not been included in this chapter, because not only is it at present a far less established application than the others mentioned, but also because it is generally expected to in fact become the major application for second generation robots. Details of this potentially very important area of robotics are dealt with in chapter 11. The remainder of this chapter deals with some of the major 'traditional' robotic workpiece-handling areas in alphabetical order.

Deburring

Robotic **deburring** is a relatively new application, as it requires a high sophistication of robot. For all but extremely long runs, the task of deburring is usually done manually, with a human inspecting machined parts for sharp metal, or sometimes plastic, irregularities (burrs) which he must then grind or cut off using a powered hand tool. Such a task is time consuming and frequently adds significantly to the unit cost of production, yet, because the shape and size of burrs for a given design of part are highly unpredictable, mechanization of the task is very problematic.

Current applications of robotic deburring tend not to employ significant sensing, although, clearly, there is great potential for sophisticated vision (or maybe taction) to inspect the workpieces and isolate the burrs. Without such sophistication it is necessary for the deburring tool to pass over the whole surface of a workpiece, whether it needs deburring or not! Because the power tool is only required to remove comparatively small amounts of material (unlike with **fettling** – dealt with later in this chapter) it is possible to actually mount the tool onto the end of the robot arm and have the workpiece held stationary. Such an approach is covered in the next chapter.

In many cases, however, it is the workpiece which is manipulated by the robot, being brought to bear against a suitable deburrer, and this approach allows different tools to be used for different parts of the deburring operation. Such a range of tooling may be necessary if, for example, there are regions which would be inaccessible if the conventional tooling was employed. Suitable robots must have sophisticated continuous-path facilities with high repeatability and a wide range of speeds. In addition it must be easy to program them with a wide variety of workpiece shapes, and to switch rapidly between the programs. Such a set-up may equally well be used for such applications as grinding the edges of glass, or even for certain metal-polishing tasks.

Diecasting

Servicing a **diecasting** machine was the original application for an industrial robot, back in 1961, and it remains a common task for robots. Diecasting of nonferrous metals involves pumping the molten metal at pressures hundreds of times higher than that of the atmosphere into moulds (usually formed from two halves) called **dies**, letting the metal solidify (sometimes aided by water cooling of the dies), ejecting the casting (usually by means of automatically extending pins build into the dies), quenching the casting in water (if necessary) and then placing it into a trimming press to remove all the excess metal (**flash**). Meanwhile, the dies must be cleaned, often by air jets, and then lubricated to prevent adhesion with the next casting.

Such an application is ideally suited to simple robots. Part orientation is precise and consistent, and gripping is comparatively simple (usually using standard grippers to grasp the stub of excess metal, called the **sprue**, formed in the channel where metal was forced into the dies). Similarly, not only does the existing equipment hardly need to be altered when robots are installed in place of human workers, but also the displaced workers may be only too glad to escape a tedious and dirty job in an unpleasant environment. The close proximity of pressurized molten metal typically involves both risk as well as hot and cramped working conditions, while the need for an operator to place his hands inside the dies exposes him to the risk of accidental closure of the dies.

Robots can be employed for any (or all) of the tasks of unloading, quenching and trimming, and typically result in improved speed, accuracy, capacity and safety, and so better utilisation of capital equipment. In some cases where components made of material other than the casting alloy must be inserted into the dies prior to the casting operation (resulting in a **cast-in insert**), robots are able to take over this task as well. For simple unloading applications the robot may even be substituted by a pick-and-place device, but in most cases point-to-point industrial robots (which must be insensitive to heat and dirt) are used to allow for more sophisticated operation.

Fettling

Despite the unpleasant working conditions in foundries, robots have made suprisingly little impact on the industry. Nevertheless, an increasingly important robotic application in this field is the cleaning (or **fettling**) of castings after they have been freed from the mould (usually made of sand) into which the molten metal was originally poured. Prior to fettling, the casting will be covered by pieces of unwanted metal: some that squeezed out around the mould, and some from the sprue, vent holes and interlinking channels of the mould. All this must be cut away, either with a flame cutter, or usually with a grinding tool, and because batch sizes in foundries tend to be low the operation must be done manually.

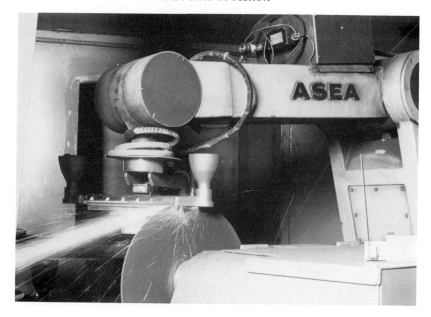

Figure 9.1 An example of a robot cleaning castings

Castings frequently have complex shapes, and their fettling is consequently a time consuming (and so expensive) task. 20% of a foundry's running costs may be accounted for by the fettling shop, and the job itself is one of the most arduous and troublesome in modern industry, with a high turnover of personnel. Distaste for the hot, noisy and dirty job has been further exacerbated by the growing knowledge of the harmful effects of vibrating hand tools on the blood vessels, nerve fibres and bones of the hands and arms. Attempts to reduce such vibration using smaller hand tools results in lower efficiency, whereas use of a robot in place of a human removes the health hazard completely while allowing flexible utilisation of larger and more efficient tools.

Owing to the large quantities of material which must sometimes be removed during fettling, it usually makes sense for a robot to handle the casting and use a fixed grinder (rather than move a grinder over a fixed casting, as possible in deburring). Such an arrangement allows the use of powerful grinders of maybe 25 kW. When the casting is held against the grinder, the robot will experience severe dynamic loading, and the arm (and the gripper) must be powerful enough to accommodate it. As with deburring, lack of the sensory feedback (inherent in a human) required to indicate where fettling must take place will result in substantial time penalties, and problems for robots may also arise from the difficulty of presenting all faces of the casting to the grinder.

Forging

As in the foundry, **forging** hot metal by means of hammer blows or presses imposes substantial physical demands on a human operator. Evolved from the old techniques of the village blacksmith, forging can involve various techniques. With **drop forging**, heated metal billets are repeatedly hit, either between a hammer and anvil in **hammer forging**, or between two dies in **die forging**. In contrast to die forging, with **press forging** the hot billet is placed between two dies and is then slowly pressed into shape in one operation. Various more specialised forging operations include **roll forging**, where billets are imparted with a required cross-section by passing them under specially shaped rollers, and **upset forging** in which, for instance, shaft-ends are enlarged by hammering. This last method is already commonly fully automated.

Because of the noise, heat, dirt, vibrations, heavy weights, polluted atmosphere and risk of accident inherent in all forging operations, use of robots would commonly be encouraged. Nevertheless, old and unpredictable equipment coupled with reliance on the speed and judgement of a human operator frequently makes introduction of robots difficult. Radical alteration of part shapes during processing commonly requires complex and heavy robot grippers to accommondate them; similarly, excessive dirt, heat, and vibration require appropriate protective shielding for the robot.

Despite such problems, positioning accuracy requirements tend not to be high, and robots are capable of transferring the hot, heavy workpieces at high speeds in hostile environments for long periods. Although a human may be faster in short bursts, the robot can outstrip him in endurance, and is being increasingly introduced for feeding of all types of forging operation. Even so, the skill exhibited by experienced forge operators during such tasks as drop forging cannot yet be matched owing to lack of sensory feedback. Human operators in such processes carefully move the struck part between each blow of the hammer and control the number and strength of the blows in order to produce the required shape in the best way. A high level of visual feedback will be required before robotic systems can replace the steadily decreasing number of humans willing to do such work.

Heat treatment

By heating and cooling certain metals (particularly steel) in a specified cycle, it is possible to modify their internal atomic structures which in turn alters their physical properties such as strength, hardness and ductility. There are four commonly employed procedures carried out with steel, each varying in the temperature the steel is heated to and the rapidity with which it is subsequently cooled. Because such heating would cause oxidation of the steel if conducted in air, it is common to exclude oxygen, either by surrounding the sample with molten salt or by excluding air from the heating furnace.

The first form of heat treatment, used for hardening, involves heating the metal until very hot and then rapidly cooling it by quenching it in a tank of water, brine or oil. With some steel alloys, however, it is sufficient merely to let them cool rapidly in air. The result of such hardening is to make the metal brittle, and **tempering** is commonly employed after hardening, in order to reduce some of this brittleness. The process involves reheating the metal to a medium temperature and then allowing it to cool in air.

A third variation is designed to maintain the overall strength of the metal but to provide it with a very hard surface. Such **surface hardening** can be accomplished either by heating just the surface, followed by quenching, or by coating the hot metal with a suitable chemical flux. In cases where metal is heated surrounded by molten salt, other chemicals may be substituted for the salt to achieve the required surface chemistry.

Finally, there is a method known as **annealing**, which is used to remove stresses from metal and generally to soften it for subsequent machining. The process basically involves heating the metal to a high temperature and then allowing it to gradually cool within the furnace, although in practice there are many subtle variations to the procedure. This and all the other methods of heat treatment are ideally suited for simple point-to-point robots. The robot arm can accurately reach into a furnace and withdraw a very hot piece of metal for appropriate cooling, and, because the cycle times of the processes tend to be relatively long, it is possible for the robot to be employed successfully for a secondary task during the periods when it would otherwise be unused.

Investment casting

Investment casting is becoming increasingly used when end products are required which need little or no finishing, for which the dimensional accuracy must be very high and which must not bear the 'join marks' left by the two halves of the moulds used in conventional casting. The procedure used involves first making a wax replica of the object required. This is usually achieved by forcing wax under pressure into a metal die of the correct shape. Because the die is only used to mould wax, it tends to have a very long life, but as with die casting, cleaning and lubricating the die between each wax casting is important.

Having created a wax replica, the next step is to 'invest' the model with a coating of **refractory material**. By continuously immersing the model in a slurry and then allowing the coat to dry, it is possible to build up a seamless mould around the replica of whatever thickness is required. Preparation of these **shell moulds** is both time consuming and tedious. Once the mould is complete, the replica round which it was formed is removed by melting the wax and letting it flow out, and then the mould itself is fired in a kiln. The actual metal casting is obtained by pouring metal into the mould, and, when it has cooled, breaking the mould to release the casting.

Simple robots are ideally suited to many of the tasks in this process. Not only can they be used for the several parts transfers involved, but they can also be employed to produce the shell moulds. Once having been taught the procedure for dipping and drying, the robot can produce uniform moulds leading to consistent quality castings. In addition, as some of the moulds produced may be too heavy for a human operator to hold for long periods, the robot is in fact able to achieve productivity superior to a human's.

Machine loading/unloading

Tending CNC automatic machine tools, such as lathes, precision grinders and milling and drilling machines, is a job increasingly being taken over by robots. As labour costs increase, it makes sense to place parts in the chuck of a machine automatically, wait until the (automatic) machining cycle is finished, remove the part, possibly inspect it, then place it in the next machine and so on. Automating any orientation and inspection

Figure 9.2 Robot loading a CNC lathe using a special 'double gripper'

functions which are otherwise accomplished manually may be problematic, but in flexible production systems where they are already
largely automated, introduction of robots is highly feasible.

The typical arrangement for a **machining cell** is to have a single robot
surrounded by a number of different CNC devices. In this way, during the
periods when the robot would otherwise be idle (because the part it had
loaded was being machined) it is possible to utilise it by servicing one of
the other devices. It is even possible to arrange for the robot to change
the tools of the CNC machines, but clearly, even without this added
complexity, because a given part may radically change in size and shape
when it is machined, robot grippers tend to be complex (and heavy).
Suitable robots must be easily reprogrammable, sufficiently repeatable
and must operate at an equivalent speed to a human. The use of robotics
in more advanced **flexible manufacturing systems** is dealt with in chapter
17.

Packing, palletising and stacking

Because robots are so good at parts handling, moving heavy objects with speed and precision, they are clearly suitable for such tasks as packing objects into containers or stacking them onto pallets. Pallets are particularly useful in the factory environment, as not only are they far easier to pick up and transport than the individual articles placed on them would be, but they also maintain the relative orientation of those objects. Robots are particularly applicable when the pattern or make-up of the objects being grouped is frequently changed, so that conventional automation would be inappropriate. Indeed, with sophisticated robotic systems it is possible to assemble together previously ungrouped objects, with the robot calculating the best arrangement for packing. Nevertheless, in most cases the robot is merely required to assemble together identical objects, although it may have to stack them onto a previously partly filled pallet.

The range of objects which can be robotically stacked in this way is very high. They can be simple boxes, packages, sacks, metal billets, glass sheets or tubing, bricks or even complex shaped components such as television tubes. Naturally, such applications spread right across industry, and they are a steadily increasing robotic market. (Indeed, in the UK, Rowntree Mackintosh have tried using robots to fill boxes with a complete selection of different types of chocolates – rather than use humans for the tedious task.) Of course, once objects have been stacked, other robots can then be used to reverse the process! Such **depalletisation** may not require any form of sensory feedback, so long as the original palletisation layout is known.

Plastic moulding

The most commonly used plastics in industry are **thermoplastics** which can be substantially softened when heated but which will harden again as before when allowed to cool. Various methods have been developed to mould such plastics to a required shape. Perhaps the most important method is **injection moulding**. This is very similar to die casting, in that polymer granules are fed through a heating mechanism and then rammed into a cooled steel mould where the plastic rapidly solidifies again. The finished product (together with extra flash which must subsequently be cut off) is removed from the separated dies by means of ejector pins.

As in die casting, the operator is at risk if he has to place his arms between the dies of the machine.

Alternatives to injection moulding are **extrusion moulding** in which softened plastic is forced through a heated die to produce the required cross-section for products such as pipes, rods and sheets. The resultant continuous extrusion is subsequently shaped and cooled by rollers and then cut into the required lengths. Various other methods of plastic forming include: **blow moulding** (suitable for constructing bottles) in which still soft extruded pipe is placed between suitable dies and is then blown up like a balloon inside the mould by compressed air, and **thermoforming** in which plastic sheet is heated until it softens and is then sucked against an appropriately shaped cooled mould.

Techniques such as injection moulding are increasingly employing robots for tasks such as the removal of mouldings (particularly those that are large and heavy), placing of special inserts into the dies prior to casting, or palletising and packaging of the mouldings. Nevertheless, removal of flash remains a problem, and although some mouldings can be trimmed using robots, perhaps because the plastic-moulding environment is not considered to be particularly unpleasant, there is not so much incentive to displace workers.

Press work

Industrial presses are used to form, shape and cut sheet metal by trapping the sheet between two appropriately machined dies. The process is heavily used for production of such parts as car and aircraft panels, but although analogous to certain processes used in forging, in press work the metal is not heated, but remains at room temperature.

Different designs of die can be used either for punching and stamping or for deformation of a sheet into a chosen shape, although in the latter, judgement must be shown as to just how much the cold metal can be stretched without tearing. In most cases the pressing operation is accomplished with a single action, resulting in a very short cycle time, yet producing products which rarely require any hand finishing after removal from the press.

Because of these short cycle times, manual loading and unloading of presses tends to be a monotonous yet fatiguing task. In addition, the edges of the sheet are often sharp and can cut the operators hands, while there is the constant risk that an operator will not fully observe stringent safety precautions and become trapped by the closing press. Although much of the lighter press work continues to be done manually, because sheets are not oriented sufficiently accurately for a robot to handle them reliably, some of the heavier and slower work is now fed by simple robots (or even pick-and-place devices). In addition to loading, some installations also employ robots for unloading from the press, perhaps to then transfer the part to a different press nearby. In these cases it is easy for orientation to be maintained for the robot's benefit.

10

Robots with tools

WRIST-MOUNTED DEVICES

This second chapter on applications is concerned with those in which the robot is not used for materials transfer, but instead acts as a moving platform for some type of tool. Unlike many of the robots in the last chapter, except for a few applications such as spot welding and glue spotting, robots with tools need to have a continuous path facility. In fact, in order to accomplish some of the tasks successfully, it is necessary to employ some form of sophisticated sensory feedback. Such second-generation robots are already beginning to make an impact in industry in such applications as arc welding.

Although almost any tool which can be wielded by a human could be attached to a robot, the constraints of sensory requirements and cost effectiveness have so far tended to result in a certain number of applications becoming far more common than all the rest. It is these applications which are dealt with alphabetically in this chapter.

Adhesive and sealant application

Modern advances in the field of adhesives have led to their increasing use as replacements for more traditional methods of fastening such as rivets, especially as more and more plastic materials have begun to be used. One of the major advantages of adhesives is that stress is distributed over a far larger area than with most other fastening techniques, yet until recently metal-to-metal bonding was not as successful as plastic-to-plastic. Recently, however, it has become possible to bond even metal surfaces which are covered in oil, and so many new applications have arisen.

Manual adhesive application can involve a brush or spray gun, a mixing gun (for two-part adhesives) or a heat gun (for hot-melt adhesives). Spraying may result in the operator having to wear a mask and be provided with an efficient ventilation system. Adhesives which need to be applied to hot surfaces may result in uncomfortable working conditions, while hot-melt adhesives will cool and others will harden if the operator is not quick enough in finishing the whole application. Robots, on the other hand, can apply many adhesives and sealants much faster than a human, consistently and untiringly. Except where only spots of adhesive are required, suitable robots need to be continuous-path, usually with five axes.

Arc welding

This is one of the most difficult tasks covered in this chapter, both for a human and a robot! The process involves using a special welding gun which causes a continuous electric spark to jump between an electrode at the end of the gun and the metal being welded. The temperature of this electric **arc** is very high and is sufficient to melt a small part of the metal where the spark lands. If the electrode of the gun is made of a similar metal, then it too will melt and help to fill in the seam that is being welded. In this case the electrode usually consists of a coil of wire stored on a drum and automatically fed into the gun, to replace continuously the tip of the electrode which melts away.

If an arc was struck in air, the metal being welded would become heavily oxidised. To avoid this, air must somehow be excluded from the immediate area of the pool of molten metal beneath the welding gun. Although it is possible to coat the surface of the constantly melting electrode with a flux material which itself will melt and so prevent oxidation, it is more common to arrange for a stream of inert gas (such as helium or argon) to flow out of the end of the gun and over the area around the arc. The inert gas prohibits any oxidation, and can be fed via flexible tubing from a gas bottle some distance from the welding site. Such MIG welding (**metal inert gas**, also sometimes known as GMAW or **gas metal arc welding**) is the most common form of industrial welding, although there is a variation known as TIG (**tungsten inert gas**), in which the electrode is made of tungsten (which is not melted by the high temperature of the spark) so that any additional metal required to fill in the weld seam must be supplied from a separate rod made of the correct type of metal.

Although arc welding is ideal for forming long gas-tight joints, it is a difficult process. There are various types of seams commonly employed, notably: **butt joints** in which two sheets are attached edge to edge (figure 10.1a), **lap joints** in which the two sheets overlap (figure 10.1b), and **fillet joints** where one sheet is joined perpendicular to another (figure 10.1c). Yet for all of these, the electrode must be held at just the right

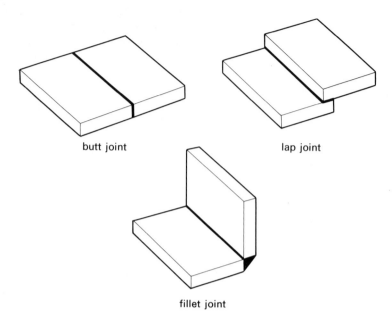

butt joint   lap joint

fillet joint

Figure 10.1 Typical types of welding joints

distance from the metal surface for the arc to melt the metal properly so
that it flows into the gap and builds up to the correct level, yet not so
close that the electrode welds itself to the surface! If the arc is held too
long in one place, the local temperature of the metal will rise too high,
and so the arc must be kept moving at an optimum rate. If the seam is
later to be exposed to pressure, the welder must try to do all of a seam in
one operation, because it is the junctions of two different seams which
tend to leak first.

In addition to concentrating on all this, the manual welder is subjected to smoke, sparks and such an intense glare from the arc that he
can only watch the process through a nearly opaque visor. It is largely
because of the unpleasantness of the job that so much interest is being
shown in using robots for welding, instead of the decreasing number of
humans willing to do it. Because of the relatively slow speeds (say
20 mm/s) and light payload involved, electrically driven robots are
commonly used, and owing to the symmetry of the gun, only five
degrees of freedom are usually required.

Teaching can either be by **lead-through**, or by teaching points which
are then interpolated by the robot. Such systems, without any form of

sensory feedback, will be able to perform welds on parts which are (and remain) accurately positioned. Indeed, they will perform far more consistently than could any human. However, in many practical applications, such positional accuracy of the parts to be welded is simply not available.

It is frequently necessary to employ some form of sensor guiding in order to compensate for inaccuracies in setting up the work, long-term wear of jigs and fixtures, thermal distortion of the workpieces during welding and **springback** of pressed components. For optimal welding of thin sheet metal such as that used for car body shells, the welding torch must be accurately positioned within 0.5 mm of the ideal position, yet commonly the workpiece itself is only positioned accurate to a few millimetres. Clearly, in order to follow the correct seam-path a robot must do more than merely mimic a taught path – it must instead use **seam tracking** technology to modify its movements. To attempt instead to reduce the positional variations of the workpieces would be prohibitively expensive.

There are two fundamentally different approaches to seam tracking. **Two-pass systems** employ a 'trial run' in which the robot passes along the expected route of the seam (at speeds up to maybe 1 m/s) without welding but monitoring, usually visually, any deviations of the actual workpieces from their expected positions. On the second (welding) run the robot is able to perform an accurate weld, usually with the sensors covered to protect them from the arc and the sparks produced. An advantage of such a system is that the speed of the sensing run can be governed by the maximum rate at which the sensor data can be processed to distinguish the seam, whereas the speed of the actual welding run can be chosen to provide an optimal weld (typically at about 10% of the sensing speed). On the other hand, such a system will be unable to compensate for any errors in the workpiece if it moves after the first run or thermally distorts during the welding process itself.

An alternative to this approach is to use a **one-pass system**. Such systems sense the seam during the actual welding process and dynamically adjust the robot's position to follow it. In the simplest systems, the robot is programmed to detect the centre of the seam, either with an electromechanical contacting sensor or else by zigzagging in a **weaving pattern** across the joint while monitoring and analysing the arc current and voltage. Such **through-the-arc sensing** systems are currently the most common form of noncontacting seam trackers. In more advanced systems the seam may be detected visually, for instance by incorporating into the weld gun a system which projects a pattern of laser light onto the workpiece surface, and then monitoring the pattern via one or more television cameras (figure 10.2). By careful filtering it is possible to prevent the glare of the arc from swamping the low-powered laser. With one-pass systems it is also necessary to protect the cameras from the spatter of the welding process with, for example, replaceable quartz windows.

One-pass tracking systems tend to be faster overall than a two-pass

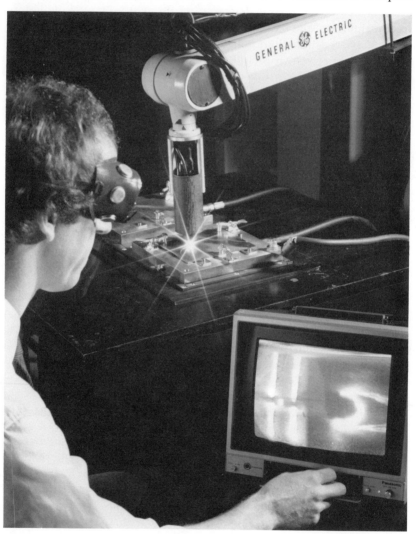

Figure 10.2 One-pass vision seam-tracking system developed by General Electric

approach, and have the advantages not only of dynamically correcting for positional errors and of potentially providing a powerful quality control technique, but also of substantially reducing the level of detail of path programming that is necessary. The possibilities of using magnetic- or acoustic-based one-pass seam tracking devices are also currently being researched, at such places as Oxford University, UK. The potential advantages of such approaches are that the data processing is simpler than with vision, the devices should be cheaper and the magnetic 'eddy current' approach is immune to fumes from the arc. Nevertheless, for complex geometries, vision is likely to remain the only option.

Lasers

Over 20 years after lasers were invented, high powered versions are increasingly being used in industry for cutting complex shapes in a wide variety of materials including metal, plastics, composites, ceramics, wood and even cloth (for bespoke tailoring). For materials such as metals the laser is usually a CO_2 type of 1 kW or more power. Sheets can be passed under the laser beam in such a way as to cut the required shape, which can be as complex or as simple as desired. There is no tool wear, and the laser can produce narrow, parallel-sided cuts without exerting any mechanical force which might distort the workpiece. In addition, there is only a very narrow heat-affected area, and the cut can be started or stopped almost instantaneously. However, in addition to cutting, the laser can also be used for welding (at faster speeds than conventional arc welding) or, by 'fanning out' the beam, for surface treatment of metals. The flexibility of the laser consequently makes it ideal for **flexible manufacturing systems**.

Clearly, many of the laser applications could be enhanced if the beam could be manipulated through five degrees of freedom to provide full three-dimensional capability, including the ability to maintain the beam perpendicular to a complex surface. However, lasers of the power required are too large and heavy to be manipulated by robot to the accuracy required. Recently however, it has become possible to obtain conventional robots which have a flexible **light guide** attached. An example of such a system is one developed by General Electric, USA, which links a 400W neodymium-yttrium aluminium garnet laser with a Hitachi robot end-effector by means of a 1 mm diameter flexible quartz optical fibre. In the past there have been problems in passing useful amounts of power down thin optical fibres, but these are increasingly being overcome. It is claimed that 300 m could separate robot and laser,

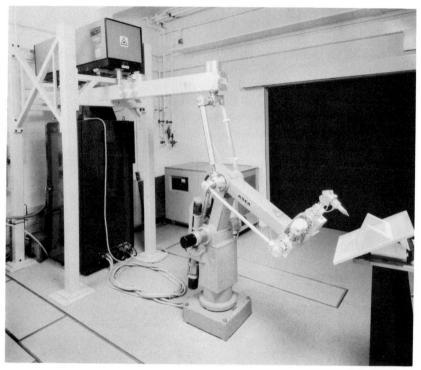

Figure 10.3 The 'Cobra' robotic laser system using a 400W laser

with only 20% loss in power. Indeed, teams of conventional robots could be linked to a single laser costing $100,000, each via an optical coupling costing only $10,000.

An alternative approach is the 'Cobra' system shown in figure 10.3. In this system a Ferranti 400W or 1250W CO_2 laser is positioned on a gantry next to a 5 DOF ASEA 'IRb6' robot arm, and the laser beam is guided to the cutting nozzle by a series of ten mirrors in articulated joints. At the final stage the beam is focussed by a lens on to the surface of the workpiece. Pressurised gas is introduced around the lens to cool it, and if the gas is inert, oxide formation is inhibited. Alternatively, blowing oxygen onto the cut can often substantially increase the maximum cutting speed. In either case, the gas blows fumes and waste away from the cutting area. Suitable robots must have a wide range of speeds, because for cutting the beam must move slowly but accurately, whereas for surface heat treatment it may have to move at up to 1 m/s.

Power tools

In addition to the various tool end-effectors employed for the specific tasks covered by the other sections in this chapter, there are also a large number of power tools (usually pneumatic) which, just as easily as being handled by a human, can instead be attached directly to a robot arm, provided that little sensory feedback is necessary for their use. It is particularly appropriate to fix a tool onto a robot (rather than use the arm to move a workpiece to a fixed tool) when the workpiece is heavier than the tool. If different versions of the tool are required for operations on the same workpiece, then a method for automatically changing tools can be employed, analogous to gripper changing. Nevertheless, the same time penalties will be incurred as with grippers, so that in these cases it may in fact be preferable for the robot to grasp the workpiece (if possible) and move it between different fixed tools.

Among the various tools which are commonly attached to robots are various forms of drill and countersink, sometimes with mechanical positioning guides to increase both accuracy and speed, and using preloading to prevent wandering. As mentioned in the last chapter, deburring can be accomplished by a grinder attached to the robot arm, and with the appropriate tooling so too can engraving. Sanding or polishing of objects (such as stainless steel sinks) can be carried out by attaching either an orbital or a belt version of the appropriate tool.

Spot welding

Throughout the 1970s the one major application for industrial robots was automotive **spot welding**. Although other applications are now increasing in importance, it remains a very large market for first-generation robots. Spot welding (also known as **resistance welding**) is suitable for bonding most kinds of metal sheet, particularly steel. The process involves clamping the sheets for about a second between two heavy-duty electrodes. A very heavy current (up to 1500 A) is briefly passed between the electrodes through the two sheets of conducting metal sandwiched in the middle. The resistance of the small column of metal directly between the two electrodes to such a high flow of current causes that local area of both sheets to melt and so fuse together in a spot.

The electrode assembly usually consists of a 'gun' of two electrodes which, rather like a pair of fire tongs, can hinge together with the sheets of metal between them. If the gun is designed to weld a long way from the edge of the sheets, it must have extra-long electrodes. In an effort to extend the electrode life, the tips are replaceable and the electrodes themselves are usually water cooled. However, such guns tend to be extremely heavy and unwieldy, even if partly supported by a suspension system from above, and in addition, the heavy-current cables must also be dragged around with the gun. Because the welding process itself is very short, it is important to move the gun to the next weld position as quickly as possible. For a human operator this is a strenuous task.

Robot spot welding is particularly attractive, not only to remove humans from the drudgery of the task, but also because a robot can get away with welding a smaller number of (accurately placed) spots than an (inaccurate) human could. Nevertheless, to accomplish the task a robot must exhibit a high degree of rigidity, and have strong drives for six axes, with rapid acceleration characteristics. Although commonly only point-to-point robots have been used, together with an indexing production line, there are increasing requirements for spot welding robots which can operate on continuously moving objects, so requiring sophisticated continuous-path devices with **full-tracking** capability.

Spray painting and coating

As with the last application, this has been a major use for first generation robots. Whenever large numbers of products must be coated a spray technique tends to be used. The surface is most commonly a fast-drying paint, but can equally well be enamel or even underbody protection. In order for a high quality to be reached, the spray gun must be kept moving and held at the correct distance from the surface being sprayed, applying a series of thin coats. Although this process can to an extent be automated using dedicated equipment (as in the automotive industry), such technology cannot successfully cope with all cases, such as points which are difficult to reach into. On such occasions humans must be used! Such a task is a skilled job, yet is also extremely unpleasant. Dust-free painting at the correct temperature requires working in small volumes, and wearing masks is vital, as many solvents are toxic and some pigments may be carcinogenic. In addition, eardefenders must be worn, as the high noise levels created by the spraying can otherwise cause irreversible damage.

Because of the need for a high level of ventilation in spray booths, a very large amount of energy is used, adding substantially to the cost of spraying. If humans can be replaced entirely from the booth, then significant cost benefits may be obtained merely by cutting down on the required ventilation. Continuous path robots taught by a skilled painter employing a lead-through technique, either using the robot arm or a **teaching arm**, are ideally suited for such automation, and various firms, such as Trallfa have built robots intended specifically for this market. Because painting is often accomplished on parts moving along a conveyor, the parts must either be guaranteed to move at a consistent rate or else the spraying robots must have a full-tracking capability.

Water jets

There is an increasing number of new applications for robot-carried tools which, although relatively uncommon at the moment, may become much more widespread. An example of these is the use of high powered water jets for both cleaning and cutting purposes. Such systems employ special pumps capable of generating a thin jet of water at intense pressure. When directed at an encrusted object one of these jets can clean the surface very rapidly. Indeed, the USA company United Space Boosters uses a robot-guided water jet to clean off charred and damaged surface layers from the Space Shuttles' reusable boosters after each flight.

With finer jets (about 0.1 mm diameter), it is actually possible to cut through a variety of materials, such as glass-reinforced plastic (figure 10.4). The jet produces a very clean cut, yet does not cause any of the dust associated with mechanical methods of cutting. By adding a suspension of specially chosen particles into the water, it is even possible to cut thin metal sheet with the process! It is relatively straightforward to attach a high pressure water jet system to the end of a robot arm, and it is likely that applications of such systems, ranging from cleaning metal to trimming motor-bike helmets, will become ever more numerous examples of 'robots with tools'. . .

Figure 10.4 Water-jet cutting and trimming of glass-reinforced composite crash helmets

11

The massive future market
ROBOTIC ASSEMBLY

Robotic industrial assembly is an application of the 1980s. Until now there were basically only two options available for assembling components. If the required product volume was large enough, it was feasible to employ dedicated **fixed automation** for the task; if the batch size was not so large, or the complexity of the assembly was too high, then the only other option was to resort to manual labour.

Although the first-generation robots of the 1960s were ideally suited for medium to small batch work, it was nevertheless soon appreciated that the assembly task required too high a level of sensory feedback and adaptability to be a suitable application, and consequently interest in building a **flexible assembly system** (FAS), sometimes referred to as a **programmable assembly system** (PAS), was not aroused again until the emergence in the late 1970s of the early second-generation robots. With the limited sensory feedback which these exhibited, together with the complex program branching which they supported, it at last became possible to experiment with robotic assembly, and increasingly industry is starting to actually apply such systems in practical work.

A 1970 census in the USA discovered that a third of car factory employees worked on assembly. Indeed it has been estimated that in many countries' manufacturing industries assembly accounts for more than 50% of the total manufacturing cost of a product, and possibly 40% of the work force. Assembly is clearly an area in which savings made by successful implementation of robotic systems would have a major impact on industry. Largely because of these potential savings a substantial degree of interest was generated in the early 1980s, as people realised that assembly seemed destined to become one of the major industrial applications for second-generation robots.

Nevertheless, there is still a long way to go before complicated tasks such as automotive assembly can be wholly delegated to robots – despite the Fiat adverts which claimed their cars were 'handbuilt by robots'! Even so, Volkswagen have recently introduced a large degree of robotic assembly for their specially redesigned Golf model, using pre-assembled

units. In this system, robots put on all four wheels, the car front
(complete with previously adjusted headlamps), the engine, piping,
some interior and so on.

Product design

The design of the components which are to be assembled is of fundamental importance in robotic assembly. Traditionally, far too great a
reliance has been placed upon the ability of a human to employ judgement, and upon the sophistication and versatility of his sensory and
manipulative skills. When a component is first designed, careful consideration is nearly always given to the type of machinery available to
manufacture it. The product is usually constrained by, and designed to
suit, that machinery, and it is only under special circumstances that
machinery is actually constructed to specifically manufacture a required
design.

   In contrast to this, in the past, products have hardly ever been
designed for ease of assembly, either by machine or human, simply
because if the assembly was above the complexity which a machine
could handle then a human could always be counted on to cope! Even
when automatic assembly devices *have* been designed, it has nearly
always been *after* the product design itself has long since been finalised.
Because of this past lack of attention to design for ease of assembly,
many consider that it is an area which, if treated as a fundamental design
stage in its own right, would be likely to result in significant assembly-
cost reductions. Indeed some have gone so far as to suggest that many of
the supposed savings due to robotic assembly may be due not to the fact
that a product was redesigned and subsequently assembled by robot, but
simply because the product was redesigned at all! For example, rede-
signing the casing of a product as one single plastic moulding, rather
than several individual 'sides' which each need bolting on, may make
assembly just as much simpler for a human as for a robot.

   Product groups which can all be assembled by similar equipment are
sometimes termed **product families**, although such families may be
radically different from those products grouped together in the company
catalogue. Typical families consist of different versions of fundamentally similar products, such as motors or watches. Some of the different
types of structures which can be distinguished in a product design are
shown in figure 11.1. These include: **frame products**, in which all components are fastened onto a frame; **stacked products**, built up from the
bottom to the top in 'pancake fashion' (and often fastened by something

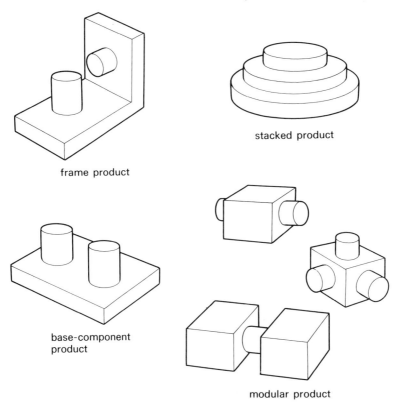

Figure 11.1 Different product design structures

as simple as a single bolt running up through the centre); **base-component products**, in which one component acts as a base during assembly and transport; together with **modular products**, in which various subassemblies are combined in different ways to produce different products.

Once the product structure has been determined, many of the assembly problems, and with them the appropriate assembly method, are already in fact frozen, so it is in practice necessary to consider detailed component design in parallel with product structure. This involves specifying the shapes, dimensions, orientations, materials, surface qualities and tolerances that are required. Surfaces are described as **functional** if they are actually employed for some purpose; they could be **connecting surfaces** which touch other components, or **assembly surfaces** which are used in the assembly process for orientation, transport, positioning or guiding. Free surfaces are described as **nonfunctional**.

In designing for robotic assembly, the number of discrete components should be reduced as far as possible. It may even be possible effectively to reduce the component count to only one, by altering the production method so that otherwise discrete parts are, in fact, integrated together when they are first manufactured. The method of using 'cast-in inserts' mentioned in chapter 9 is often a suitable approach for such component integration. Similarly, because orientation is usually so expensive, some of the methods covered in chapter 8 such as magazining, palletising, using component bands or linked components or integrating production into the assembly process, may be used instead, although these too can be expensive. If orientation is still necessary, attempts should be made to simplify the process by avoiding components of low quality, and by making sure that those used are either symmetrical or else clearly asymmetrical (possibly by adding a special feature to the part).

Parts-feeding problems can be alleviated by avoiding component designs which will ride up on to each other, wedge or tangle. For example, it would be advantageous to replace loosely coiled open-ended helical springs with tightly coiled springs with closed ends. In addition, feeding costs may be reduced if two components can be designed with a sufficiently similar form that they can both be fed using the same equipment. Transport of the assembly itself is eased if some form of base component is available.

An early study which analysed a small range of different nonrobotically assembled products discovered that in most cases over half of the components were assembled from one dominant direction, about 20%, more came from the opposite direction, and a further 10% arrived in a plane perpendicular to these two. Only the remaining 10% of the parts arrived from more complex directions, suggesting that, traditionally, products are predominantly stacks.

As a result of work such as this, it is generally thought that robotic assembly should ideally be in layer fashion from above along a vertical axis, or at least from as few directions as possible. This may allow the assembly to be accomplished using less than the six degrees of freedom that would be expected, and consequently the robot required may be far cheaper. In practice, however, many have found that the possible problems caused by parts being slightly out of alignment make the restrictions imposed by robots of less than 6 DOF unnecessarily limiting, so outweighing any cost advantages. Nevertheless, whatever approach is taken, it is of course paramount that the product design should provide the robot arm with unobstructed access!

To increase the ease with which difficult parts can be picked up by the robot gripper, special features may be added, such as a hole which can self-centre the part if an end-effector opens up inside the hole for an internal grip), or even just a flat surface which would enable the part to be lifted using a vacuum cup. The same analysis of sample products referred to above also distinguished various 'assembly tasks', such as 'insertion of peg-like objects', which was found to be the most common task, and was usually along the dominant approach direction. The next most common task was screw insertion, although this was typically from a perpendicular direction. It was found that all of the surveyed products could be assembled using combinations of these and ten other operations: welding or soldering, crimping, providing temporary support, removing temporary support, product inversion, locating-pin removal, force fitting, insertion of a peg-like component and retainer, insertion of a multiple-peg-like component and, finally, push-and-twist of a component.

Of these tasks, it may be possible to avoid the fastening techniques by using integrating production methods. When this is infeasible, the technique employed should be chosen with care, bearing in mind the relative difficulties of the different methods shown in figure 11.2. Any insertion can be aided with chamfers on one or both of the parts, and whenever possible, self-aligning and self-fixturing parts (for instance, pins designed into one component with corresponding holes in another) should be employed. Snap-together parts may occasionally be feasible, but frequently the product must be capable of subsequent disassembly, and so screws may be the only practical solution. When this is so, the number, types and sizes of the screws should be kept to a minimum.

EASY FOR A ROBOT

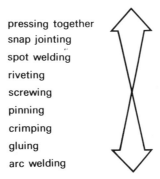

pressing together
snap jointing
spot welding
riveting
screwing
pinning
crimping
gluing
arc welding

DIFFICULT FOR A ROBOT

Figure 11.2 The relative difficulties for a robot of different fastening techniques

Assembly sequencing

The components of a given product design can typically be assembled in a very wide variety of different sequences. In addition, the sequence best suited to manual assembly may not be the optimum when robots are employed. For instance, with a multirobot FAS there is a need carefully to 'balance' the operations each robot performs in order to reduce the time any one part of the system is idle. As a result, it is necessary somehow to select from all of the potential assembly sequences, the one which, according to selected criteria, appears optimum. For all but very simple assemblies, it is not feasible to attempt to evaluate any but a small subset of the possible sequences, yet selecting that subset is itself often a daunting task.

The number of inversions of a product during assembly should be reduced to the minimum because this task is not only difficult but also time consuming. As explained in the context of grippers (chapter 8), with robotic assembly any approach which saves the robot time tends to have an immediate saving in unit cost, because it is usually the 'slow' robot which dictates the speed of the whole assembly process. Consequently, assembly sequences should be chosen which minimise the handling times of components, and if different grippers must be used, then the sequence must be designed to reduce the number of gripper changeovers. This can be accomplished either by assembling at one time all the product components requiring a given gripper or, if this is impossible, by assembling a small batch of products in parallel so that the time wasted by changeover is spread over several products.

Comparisons of different assembly sequences require some method of formalised description. One approach to this, developed at the Charles Stark Draper Laboratory, USA is to employ **parts trees**, of which a very simple example is shown in figure 11.3, which describes the construction of a 5-component product. Construction of a tree involves first listing all of the components together with an 'orientation code' which indicates the alignment of the component necessary for correct assembly. In the example shown, component D might be a bolt which had to be assembled with its head uppermost, corresponding to a code of '1'. Components such as nuts and washers are symmetrical, so an orientation code is not required (for instance, component E).

With the components listed across the top of the parts-tree the time-axis is considered to run from the top to the bottom. Addition of a component or **sub-assembly** (considered to be a collection of components which will not fall apart when moved or inverted) to the main

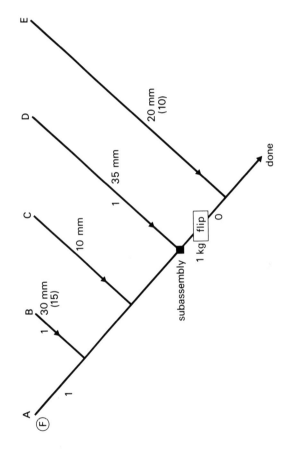

Figure 11.3 A simple example of a parts tree

assembly is symbolised by a line with an arrow, and the assembly actually takes place where lines meet at a node. Lines between components which are already joined do not have an arrow. Special fixtures (which should be avoided if possible) are signified by an 'F', while any special operations such as parts inversion are specifically indicated (in the case of inversion the weight of the subassembly is also stated).

The orientation code is placed close to each arrow (if necessary), together with the distance in millimetres of the shortest path from the component to the assembly. Some parts can be dropped into the assembly from a certain distance away, and in these cases the second distance is placed in parentheses. From the distances included it is possible to estimate the time that would be taken for each movement, and from these to estimate the time for the whole assembly.

Using such assembly-sequence descriptions it is possible to try to evaluate different approaches and by recognising shortcomings of a given sequence try to overcome them by modifying the assembly. A sequence is usually considered preferable if it contains fewer inversion operations, special fixtures and subassemblies, but there is an unclear trade-off between speed, cost and complexity. At present, even the design of the different assembly sequences which are to be compared involves the creative insight of a human designer, although (unsuccessful) attempts have been made to use graph theory and linear programming techniques to generate likely designs automatically.

As computing power increases it is probable that such approaches will eventually be fruitful, with optimising programs choosing the very best assembly sequence for a given product. Nevertheless, it seems likely that, because of the large number of potential sequences, even advanced programs of this type will need to employ various **heuristics** or 'rules of thumb' to narrow down the search, rather than attempt to evaluate all of the possible choices.

Layout, analysis and evaluation

As with assembly-sequence selection, deciding with any certainty on the optimum physical layout for an FAS is impossible. With careful design work and subsequent analysis it may be possible to arrive at a layout which out-performs any existing method of assembling the product, allowing one to say that the system is 'better'. Without any standard against which to measure this performance, however, it is hard to say that the system is the 'best' that could be achieved.

Somehow, the FAS design process must reflect the desired product-design flexibility, quality and volume requirements in its choice of robot or robots, feeders, conveyors, orientation systems, inspection devices and overall geometric layout. Yet on the one hand, clearly, any 'brute force' attempt to compute every possible configuration is doomed to failure, while on the other, the simplifications and approximations required for an **operational research** (OR) type approach have, up till now, tended to result in layouts which are technically not even feasible. Of all the stages in the design of robotic assembly, this is probably the most difficult and most prone to errors.

Testing of each proposed layout involves conducting some form of cycle-time analysis to obtain the estimated production time for the assembly, followed by a cost analysis to indicate whether the projected robotic assembly would be cost effective. As a very rough rule of thumb, a typical one-component assembly cycle tends to be about five seconds or longer. In general, anything faster than this is currently likely to be too rapid for a robotic approach. Various methods of cycle-time analysis have been suggested for robotics, all basically making use of the fact that if the dynamics and control algorithms of a robot are known it should be relatively straightforward to calculate assembly times.

Methods Time Measurement (MTM) is a methodology established in the 1940s for human work analysis, and with it an industrial task can be broken down into elementary motions, and the time for each motion and the total estimated time for the task can be derived. Workers at Purdue University have developed a methodology analagous to MTM termed **robot time and motion** (RTM). They have had to define new task motion elements for robots because many MTM elements aggregate basic robot motions, while others are impossible for a robot to perform. Given a task described in MTM it is possible to translate it directly into RTM, although many elements will translate into impossible tasks for a robot, indicating that the work method must be changed. Consequently, although a list of RTM elements will yield an estimate of robot cycle time (and a computer can even then convert the RTM into an actual program to run on the robot concerned) the initial design of the RTM program frequently remains problematic.

Cost analysis and justification for robotic assembly, and indeed for robotics in general, is in bad need of rationalisation. The subject is dealt with in detail in chapter 16. Nevertheless, some basic work on the justification of robot assembly in comparison to existing methods has been carried out using specially developed economic models. The simplified example illustrated in figure 11.4, shows robotic assembly to be

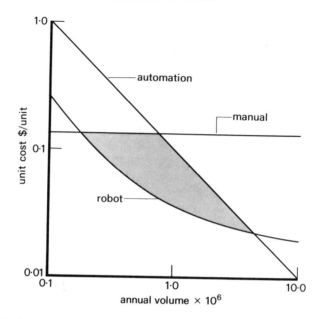

Figure 11.4 Regions of economic advantage for three different assembly methods

the most cost-effective choice in the medium-volume region. Extensions to this work have suggested similar trends, largely dependent on various aspects of assembly such as quality of parts and frequencies of design changes.

Optimal robotic assembly

Because of all the design difficulties inherent in constructing an FAS, it is so tedious to 'rerun' through the whole procedure that the methodology is far from sensitive to changes of design of product or robot. Instead, all that tends to occur is that attempts are made to 'modify' the existing FAS if gross changes demand it. For a technology specifically designed for flexibility, this is a great shortcoming. Work conducted at Imperial College, and outlined below, attempts to suggest a robust yet sensitive methodology which can be employed when investigating the cost effectiveness of assembly by robot.

When considering robotic assembly it is sometimes convenient to treat the costs of the flexible, reusable portions of the system separately from the costs of those parts of the expenditure (including design costs) which are 'dedicated' or 'fixed' to a particular product design and are of no further benefit when that product type is no longer being assembled. This **principle of segregated cost rates** immediately highlights the answer to the question: 'Is it better to use a cheaper or a faster robot – which has most effect on the unit cost?' Because unit cost of assembly is equal to the product of the cycle time and the joint cost rates of the flexible and the fixed portions of the system, that is:

$$\text{Unit cost} = \text{cycle time} + \left(\begin{array}{cc} \text{cost per second} & \text{cost per second} \\ \text{of the flexible} & + & \text{of the fixed} \\ \text{part of the FAS} & \text{part of the FAS} \end{array} \right)$$

it is better to use a robot which is twice as fast rather than half as expensive – a halved cycle time is multiplied by the whole of the brackets, whereas a halved robot cost-rate is just one of the terms within the brackets.

Similarly, it becomes clear that as the numbers of *different* products to be assembled increases, so the amount of dedicated equipment and costs fixed to a given product should be reduced, and the amount spent on flexible equipment (common to all batches) increased to compensate. However, because it is usually quicker (and so cheaper for large runs) to assemble using dedicated automation, there is a balance to be maintained, illustrated in figure 11.5, depending upon how frequent the

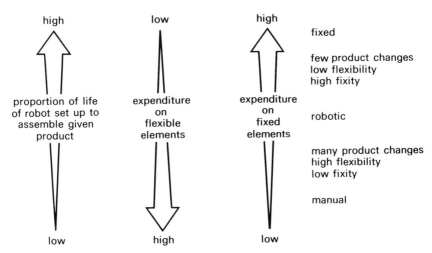

Figure 11.5 The altering balance of the fixed and flexible elements of an 'ideal' system depending on the frequency of changes of products assembled

design changes are to be. Towards the extreme, for very long run lengths, what little flexibility is exhibited by the system is obtained by using special 'fixed' automation which can in fact physically be altered to allow assembly of any particular product from a small product family. For still longer runs the system used is of the traditional 'hard automation' type. This concept is crystallised in the **principle of fixity** which states that:

> For optimal robotic assembly, the proportion of the total cost of an FAS which is due to expenditure 'fixed' to a particular product is directly related to the total time spent throughout the life of the system assembling that given product.

Upon investigation of the above it becomes clear that there is an implicit **principle of specific run-length** in a given FAS design which makes it best suited for assembling different products, all for approximately the same total length of time (for instance, a new design every six months). If an FAS, once built (and the robot(s) bought, so that the 'flexibility' cannot be changed), then has to adapt to a shorter run length than originally designed for, figure 11.6a, the robot may be hard pressed to compensate for the necessary loss of dedicated equipment imposed by maintaining the correct **fixity ratio** (of 'fixed' expenditure to the whole). Similarly, if the run length is made longer than the FAS was originally 'tuned' for (figure 11.6b), maintenance of the correct fixity ratio will result in an oversophisticated system.

In actually choosing a suitable robot for a given assembly task, the most appropriate choice is not necessarily the least sophisticated or least expensive model which is still capable of accomplishing the task, but may instead be that with the lowest ratio of cost-rate to speed. In other words, a more expensive robot may be capable of performing the assembly so much faster that, provided the increased outlay can be covered, it is a preferable option to a less advanced model. Clearly, in many cases more than one robot arm is in fact required, and there are various distinct types of such 'upgrading'.

The simplest (although rarely the most appropriate) approach is simply to duplicate the whole of a single-arm system. This obviously involves a larger financial outlay, but it reduces system cycle-time in inverse proportion to the level of duplication. Unit cost of assembly is only affected by such factors as shorter assembly times incurring reduced overheads, multiple systems being purchased at a discount or significant initial design costs not needing to be duplicated.

An alternative to using multirobot systems is to use a single robot with

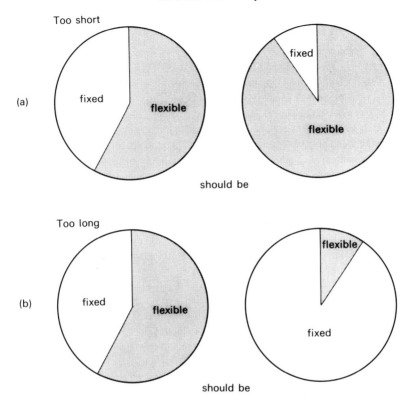

Figure 11.6 Illustration of the principle of specific run-length

more than one arm which divides the assembly task into a series of parallel tasks. Because a multiarm robot tends to be cheaper than the equivalent number of individual single arm robots, this approach is usually preferable to a 'duplicated' system. Use of different sophistications of single arm robots reduces cycle time but not as much as the previous two approaches. In this case unit cost of assembly is only reduced if the lower cost-rates of the less sophisticated robots counteract the effects of the probable incomplete utilisation of the overall system. In choosing between these last two systems it is felt that the decision to opt for a multiarm system should be determined by the complexity of the assembly, whereas choosing a multirobot system is a function not of the complexity but of the *range* of complexities within the assembly.

If care is taken when applying such upgrades, then first evaluation of a robot assembly system can be made for a design employing only one or more identical single-arm robots, knowing that any subsequent practical designs will exhibit a similar or an improved performance. Detailed FAS design remains a complex business and much research, some mentioned in chapter 17, is currently being funded to improve our understanding of the very wide ranging factors involved in the assembly process we are trying to robotise. Even before such work is completed, industrial applications will continue to grow rapidly. To the modern industrialist, understanding the detailed principles of assembly may for the moment remain something of a luxury. For him, building an (albeit unoptimised) robotic batch-assembly system may soon be an act of survival, rather than of tentative academic interest.

12

Science fact from science fiction
MOBILE AND OTHER ROBOT-RELATED DEVICES

Autonomy versus teleoperation

This chapter deals with what are, for many, the 'real' robots: robots which are mobile, have senses, a limited degree of 'intelligence' and possibly even anthropomorphic limbs. These are the 'Star Wars' robots, the robots of (until recently) only our imagination. They are the designs which the man in the street still tends to think of when the word 'robot' is mentioned, yet they are also the designs which many believe remain in the realms of science fiction. To an extent they still do, but already there are many mobile robots in research laboratories (also dealt with in chapter 17), as well as several examples of 'eye-catching' devices which are closely related to robots (and certainly to robotics) and are commercially available. In some countries these devices are thought of as actually *being* robots.

As explained in chapter 1, not all robot-like devices are, in the West, correctly termed 'robots', because they do not have the autonomy of control which is necessary to be a true robot, but instead are controlled remotely by a human. Such **teleoperators** (or **telechiric devices** or **telechirs**, from the Greek 'distant hands') include any manipulation system which is remotely controlled by a human operator in response to sensory information transmitted from the workplace. The term therefore includes the purely mechanical **master-slave manipulators** used for handling radioactive material (figure 12.1), originally developed during the Second World War, where force feedback is through direct mechanical linkage. In 1947, the first servoed electric-powered teleoperator was developed, but no force feedback was available for the operator. A year later, however, a new system was developed in which force information was relayed to the operator by backdriving the master. Few would call purely mechanical teleoperators 'robots', yet when the linkage ceases to be mechanical, but instead becomes totally

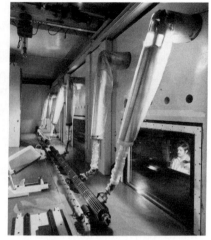

Figure 12.1 Examples of mechanical master-slave manipulators

electrical, then to many the device starts to seem very similar to a robot.

To a large extent this is a valid view, for the actual hardware of the teleoperator may be identical to that of an advanced robot; indeed, all that may need to be done in order to convert the device to a 'robot' is to replace the human by a computer system (as was done in 1961 at MIT's Lincoln Laboratory). The physical remoteness of such a computer in no way alters the device's right to the term 'robot' (although it would be more usual for the computer to be actually situated with the rest of the device). In practice, a teleoperator system may often be a convenient intermediate stage in a developing design, for although sufficient computing sophistication may in many cases not yet be available, several mobile-robot applications ideally require at least a certain level of autonomous control because of the inherent limitations of otherwise using a human.

Not only is it increasingly expensive to pay for a skilled human to control a teleoperator full-time, but the human's performance is itself limited anyway. He can only effectively control two manipulators at once, he has slow reactions, can only concentrate on one task at any given time, and the level of mental complexity with which he can cope is comparatively low. In addition, as robots become increasingly employed in such alien environments as occur with marine and space exploration, then the control required tends to become counter-intuitive and a human operator must be given substantial training before he can operate the system satisfactorily.

In addition to such problems there are also technical limitations to teleoperator systems. There is a limited rate at which information can be successfully transferred between two given locations. Even using high-**bandwidth** transmission systems such as microwave dishes or **fibre-optic** cables, the available datarate is rapidly used up when feedback such as high-resolution colour television images must be sent from teleoperator to controller. There may also be interference problems with such systems, ranging from general 'noise' to complete blackout (if a space robot, for instance, is occluded by some spatial body).

A problem that becomes paramount when considering space applications is the question of communication time delay. One of the reasons for using space systems is to avoid the need for a human to actually be in space. The much publicised Space Shuttle teleoperator is used by a crew that are already in space anyway, but even so it has been estimated that to merely maintain a human in space for a year costs several million pounds. Consequently, control of most future teleoperators is likely to be from Earth (or some equivalent station) a significant distance away. Yet the speed of radio transmission, fast though it is, results in 0.3 seconds round-trip delay from Earth orbit, 2.6 seconds from the Moon and 10–40 minutes delay from Mars (depending on the planet's position). Studies suggest that only one tenth of a second is the maximum tolerable feedback delay for a human controlling a teleoperator in real time, implying that about 15,000 km is the maximum distance over which such a system can satisfactorily be used.

Clearly, it is possible for the human operator under such circumstances to employ a 'move-and-wait' approach, but control of a Mars 'robot' using such a strategy would result in it only performing useful functions for 5% of the time – the rest would be spent waiting! This is obviously completely inappropriate, if only because unforeseen hazards may demand rapid response. When a landslide is sensed heading for the robot, it *must* act autonomously if disaster is to be avoided. Any signal from Earth would be far too late. Because of all the above shortcomings, teleoperators are rarely seen as a final solution, and so although many of the 'robot' systems subsequently covered in this chapter do not in fact yet have full autonomy (due to insufficient computer sophistication) they are all potentially fully robotic, and as such deserve discussion.

Terrestrial robots

With wheel and tracks

Many simple automated guided vehicles have been in use in industry for several years, where they have found increasing success for warehousing and distribution tasks. Typical systems, such as those shown in figure 12.2, follow a route predetermined by laying a wire into a shallow groove cut into the floor and then filling the channel flush with the floor surface again. A frequency generator supplies a current to the wire and so creates a magnetic field which the AGV can sense. Through detecting the field by two sensors either side of the track, it is possible for the AGV to sense if it is steering off course, because one of the sensors will receive a stronger signal than the other. Consequently, correcting action can be initiated.

Such simple AGV systems can nevertheless incorporate routes with several branches, loops and spurs, by using different frequencies for each path. Programs to specify the route for a given AGV to follow can be input using a keyboard on the vehicle or else can be relayed through transmitter stations in the floor from either a remote manual control

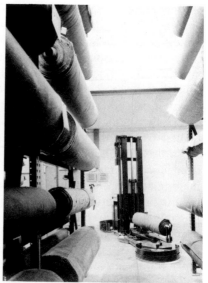

Figure 12.2   (a) Example of a typical wire-following automated guided
            vehicle
          (b) An AGV used for storing carpet rolls

station or a central control computer. In this way materials handling can be accomplished in a far more flexible fashion than with conveyor systems. Existing doors and yard and road crossings can be incorporated into the travel route, trailers can be automatically coupled if necessary, the AGV can have fork-lifting and other automatic materials-transferring facilities, or can act as a form of mobile assembly platform with programmed height adjustment for the workpiece.

An AGV with a difference has been in use since the early 1980s in various office buildings. This driverless Mailmobile is used to replace the office messenger. It follows a trail of fuorescent material round the corridors, and stops at predetermined places, bleeping to summon office staff to collect their mail before it continues on its way. Supposedly this bleeping so enraged one office worker that he pushed the machine into a lift and sent it to a distant floor which had no fluorescent tracks for it to follow, hoping no doubt that it would never return. An equally unfortunate reception occurred at the Pentagon, where it was thought that the miles of corridors would be ideal for the 'robot postman'. It took three months before somebody realised that there was not really anything to stop someone from quietly looking at all the classified documents placed on the trolley. The Mailmobile was removed. . .

All AGV must contain sufficient sensing to avoid any obstacles (including humans!) which may be placed in their path. If such sensing is expanded to include navigation aids, and the whole system is placed under the control of an on-board computer, then it is possible to create a truly free-roving vehicle which is not restricted to following predetermined routes. Such AGV are already being developed in the form of driverless fork-lift trucks. However, as soon as such a high level of disorder can be tolerated as occurs in such unstructured surroundings, suddenly mobile robots become applicable outside the relatively controlled world of the factory.

Domestic robots, capable of taking on a substantial proportion of the drudgery of housework, are still a long way off: washing dishes and making beds are suprisingly complex tasks. Vacuuming the floor (without running over the sleeping cat or knocking over the Ming vase) is possibly a lot closer, as, in principle, the problems are similar to those found in navigating through a factory or warehouse. Despite all these problems, various 'domestic robots' (at least in name) have been built, and a few are already commercially available.

They range in type from 'home personal robots' suitable for security, games, and very simple vacuuming, to 'educational robots' (such as that shown in figure 12.3) designed for 'education, entertainment, experimentation, and advertising at exhibitions'. The story goes that when one

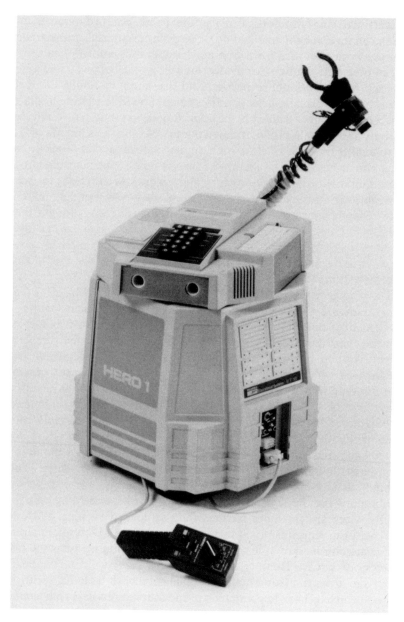

Figure 12.3 'Hero' – a mobile educational robot

of the latter robots had just been developed, and only one prototype existed, its makers had to take it on flights as hand luggage. When customs officials showed an understandable interest in the large box they were carrying and opened it up, the robot was programmed to sense the change in light level, trundle out of the container, and say in a synthesised voice, 'It sure feels good to get out of that box!'

A remote-controlled device known as 'the wheelbarrow', has long been used for the remote inspection and disarming of bombs. Later, with a robot arm placed on top, and renamed 'MR (Mobile Robot) Bill', the design became a research tool for mobile robotics at Warwick University. Eventually truly robotic bomb-disposal systems may become available, capable of traversing difficult terrain and at last able to replace the 'wheelbarrow'. Such robots may well use a track-laying system as did the original, but for very uneven ground such an approach is infeasible. In cases such as these, the robots of the not too distant future are likely not to glide to their work, but to walk. . .

Terrestrial robots with legs

Adaptive walking machines are vehicles which replace conventional tracks or wheels with articulated systems of levers. Each joint of such a system is individually powered and coordinated by some central controller. It is only recently that the increased sophistication of available computing has meant that it is becoming feasible for the complex control task to be managed automatically. Even so, the required control algorithms are very complex, requiring as they do varying gaits depending on the type of terrain and speed of travel. All this ideally requires substantial sensory feedback. An idea of the advanced nature of the task can be gathered by shutting one's eyes and trying to first walk and then run upstairs without the benefit of visual feedback; impair one's sophisticated balancing systems with a little alcohol and then try. . .

In practice, few attempts have been made to construct bipedal devices (although WABOT, an experimental anthropomorphic robot built at Waseda University, Tokyo, is a notable exception – figure 12.4). Four-legged devices are far more stable, while six-legged versions are still better because, in principle although walking they can still maintain a stable triangular base of at least three legs on the ground. Just as the old milking stools always had only three legs so that they would not 'wobble' on uneven ground, likewise the three-legged base obtained with a hexapod vehicle is usually preferable to the four-legged base that would be obtained from an eight-legged device.

Consequently, the most appropriate design seems to be to use six

Figure 12.4 'Wabot' – the Waseda University biped robot

legs. Notwithstanding this, various research work at Carnegie-Mellon University, USA, designed to investigate the detailed problems of walking, has, in fact, been conducted using a *one* legged device which constantly has to balance itself and move by hopping! Another variable in legged devices, footsize, tends to be a tradeoff between large feet providing low footprint pressure, and small feet offering good purchase.

Figure 12.5 The General Electric 'Quadruped Transporter'

Figure 12.6 The Odetics 'functionoid' lifting the back of a 2 tonne truck
and moving it round through 90°

Research into such truly robotic walking devices is still being conducted, but as far back as 1967 General Electric demonstrated the
Quadruped Transporter shown in figure 12.5. It possessed four independently moveable hydraulic legs, and was completely controlled by the
operator mounted in the control cabin. Although the device demonstrated the dramatic increase in mobility over uneven terrain obtained by
using legged transport, the manual control of the transporter was found
to be extremely arduous, even over short periods.

Recently, Odetics Inc. have developed what they call a 'functionoid'
(resisting the temptation to call it a mobile robot), shown in figure 12.6,
which walks under remote control rather like a spider, but can lift and
carry several times its own weight. Autonomous control for such devices
is likely to become available in the mid-1980s, so that, in addition to
working in hazardous environments, legged robots can perhaps at last
start to compete with the common mule, a beast which can traverse
most terrains while carrying 200 kg over 25 km in a day, or 100 kg over
50km.

Marine robots

Those industries in which man must somehow be subjected to an alien (and so dangerous) environment have all tended to be active in the development of **remotely operated vehicles** (ROV). Since the 1960s, the US Navy has been a leader in the development of undersea systems, producing such vehicles as the Remote Unmanned Work System (RUWS) and various Cable-controlled Underwater Recovery Vehicle (CURV) systems. ROV have now become a common tool for oil companies working in the North Sea as they can assist (and sometimes replace) human divers. Similarly, NASA employs ROV to retrieve the solid rocket boosters dropped into the sea by the Space Shuttle, and the nuclear industry uses them for inspection of reactors (which would otherwise have to be performed by special divers).

ROV can be broadly classified into three categories. The least sophisticated range consists of those devices designed solely for *inspection* purposes. Systems capable only of external inspection of an object will not need any form of manipulator, but those required to perform inspection from locations which the ROV itself is too large to reach will have to employ some form of manoeuvrable arm with a camera on its end. More sophisticated devices are designed for *recovery* work. This covers everything from picking up small objects to salvaging from the ocean bed. More complex recovery operations may involve the rigging of slings or lines around the desired object, which may require the use of two manipulators, but provided that the ROV is not primarily designed for making mechanical modifications to an object, it is still classified as a recovery device.

The most sophisticated form of ROV is one designed to perform actual *mechanical modifications* to an object. Such systems require two or more manipulators, maybe with some form of force-feedback measurement. When operation is totally controlled by a human, such force-feedback information (if considered absolutely necessary) can be conveyed to the operator visually. In order to perform any useful undersea work, it is usually necessary (unless the vehicle is on the ocean floor) somehow to latch the ROV on to the object of interest, so that its relative position is constant. Clearly, the design of suitable manipulators for these ROV can be problematic, as all position sensors and actuators must somehow be protected from the high-pressure salt-water medium in which they must operate.

An example of one of the more sophisticated types of ROV is Shell UK's 35 tonne Remote Maintenance Vehicle (RMV). It is the size of a double-decker bus and is designed to change valves, replace control modules and perform routine inspection in waters too deep for conventional production systems. The RMV winches itself down from a buoy and then latches onto a track which passes through the required work area. It then propels itself along the track and performs maintenance operations controlled from above via an umbilical cord.

Increasingly, control of ROVS is being automated. Currents can be automatically compensated for, certain manipulator sequences (such as tool change) can be placed under computer control, and various techniques can be used to reduce operator fatigue. Work is already far advanced in developing free swimming versions, no longer attached to the mothership by means of an umbilical but sending and receiving lower levels of information using the water as a medium. Such systems are, of necessity, highly robotic, and as advanced vision systems become available, so the vehicles may approach complete autonomy.

Space robots

There are many activities in space which can potentially be performed better by a robot than an astronaut on a spacewalk and, of course, with no risk to human life. Deep space probes, although requiring no manipulation, nevertheless may need to release a monitoring device into a planet's atmosphere at an appropriate moment (as the Galileo Jupiter probe scheduled to fly in 1985 is designed to). Because of the very long communication time-delays with Earth, short-term decisions must be made by the spacecraft itself. Similarly, scientific data collection, such as monitoring of stars or planets at different frequencies in the electromagnetic spectrum may largely need to be directed autonomously if transient phenomena are to be captured. If a landing probe is to examine samples from a planet's surface, as did Viking II which landed on Mars in 1976, great time savings can be obtained if tasks can be robotised.

A major use for space robots is seen to be in satellite servicing – including deployment, maintenance, repair and retrieval. The advanced teleoperator in the Space Shuttle is an early approach to this problem. It not only permits an astronaut to manipulate heavy objects in the open hold and in space, while still in the relative comfort of the shuttle, but sophisticated electronic feedback systems provide force-feedback to the 'master' arm being operated by the astronaut. Even simple repair of satellites can be of great benefit: in the past expensive systems have been put out of action by the blowing of a fuse. . .

Autonomous roving surface vehicles (a robotic version of the Lunar Rover used in Apollo 15 and 16) would allow large areas of a moon or planet to be surveyed in the course of a mission, and indeed the Soviet Lunokhod mission has already explored a 30 hectare area of the moon using an unmanned teleoperated eight-wheeled rover. Space construction, including assembly of such structures as space stations, solar power satelites, large antennas, or space manufacturing plants is considered to be a major application for space robots. Space manufacturing is of advantage because objects require no support and there is no convection of gases or liquids, allowing extremely pure materials to be created. Finally robots are seen as suitable for space rescue of missions such as Apollo 13 (which was aborted due to an onboard explosion) and the Skylab Space Station which was prematurely lost through reentry.

NASA is currently very active in developing space robots, spending about $3 million a year on the work. They plan to launch a remote orbital servicing system in the late 1980s. The spacecraft includes two 6 DOF manipulators, controlled from the shuttle or the ground using a sophisticated teleoperator system. Although this system is not truly robotic, NASA expects genuine space robots to become available during the 1990s which are capable of intelligent autonomous operation, working unattended in space, on tasks such as repairing satellites. The robots would be given high-level instructions, leaving to inbuilt intelligence the task of formulating exactly how to carry out the job.

Bionics

In 1960, the US Air Force held the first conference on **bionics** (BIOLOGICAL electroNICS). The term was invented by Dr Hans Oestreicher to refer to the creation of novel technological devices according to 'design principles' observed in biological organisms. Since then the term has gained widespread acceptance though its use in such television series as 'The Six Million Dollar Man'. In reality, although **prostheses** (artificial replacements for parts of the human body) have not reached the level of sophistication portrayed in such programmes, some are nevertheless becoming very advanced.

Prosthetic limbs, in particular, are becoming increasingly sophisticated and lifelike and artificial hands are commonly contolled merely by the will to move them. Electrical signals from the stump muscles are detected and cause the appropriate actuator in the prosthesis to operate. The more the stump muscle is contracted, the faster the prosthesis is made to move and the stronger its grip. In some, strain gauges in the hand feed back a proportional signal to the amputee's medial nerve, and the level of tingling he senses is proportional to the grip he is exerting. Although such systems are in no way robotic, clearly the similarity of the technology involved in both bionics and robotics makes them closely allied fields.

A similar area to prosthetics is **orthotics**, literally 'correction of crooked parts'. Walking machines are orthoses which, like artificial limbs, are becoming increasingly 'robotic-like'. Experimental **exoskeletons** have been developed which could eventually allow paraplegics to walk. Such systems can only operate well below the level of a normal human, but in the past various prototype exoskeletons have been built to try to amplify a healthy human's existing power. One of the most publicised examples was the General Electric 'Hardiman'. It was built in 1966 for tasks like warehousing and factory operations, salvaging, bomb loading and work in remote areas. Strapped into the exoskeleton, a human was able to lift several hundred kilograms without effort, yet problems with control prevented the system's adoption.

Perhaps one area of bionics which is likely to link up finally with the control theory of robots is some work being done by Dr Jarrold Petrofsky at Wright State University in Dayton, Ohio. Petrofsky has been successfully experimenting with restoring the use of paralysed limbs by using microprocessors to stimulate motor nerves in the limbs artificially (attempts to stimulate muscles directly require higher voltages, and electrodes placed on the skin then tend to burn and irritate). By analysing the pattern of impulses and muscle activities that produce a given movement in the body, it has been possible to translate the information into a computer program which can in turn stimulate a paralysed limb.

In late 1982, a paralysed human volunteer took a halting step forward using the new technique. Clearly, for someone to actually walk freely using this approach requires the limbs to have sensors attached to the joints in the same way as the equivalent bipedal robot would. The control problems are also very similar. Slowly, the boundaries and distinctions between purely biological and solely mechanical systems are being tested, and it is becoming clear that in reality such boundaries do not actually exist.

Part IV
SOCIAL, ORGANISATIONAL AND ECONOMIC CONSIDERATIONS

This very important section deals primarily with the social-science and management-science considerations of robotics. The four chapters range from a discussion of some of the various views held on the implications of robotics for both society and the individual, to an investigation of the management processes involved in robotisation. These topics are truly of just as much importance (although less immediacy) to engineers as to managers – without an appreciation of the managerial considerations tied up with 'real-world' robotics, or of the fears many feel about the direction in which robotics seems to be going, it is truly difficult to understand many of the complications which arise in the diffusion of robots.

13

'What about my job?'

ROBOTICS AND PEOPLE

Worries

There has been a lot of rubbish talked about robots and unemployment, not least by roboticists themselves! It is of little advantage for robot manufacturers to stress the possible employment problems; management are frequently concerned that their workforce are already all too aware of the problems, and researchers may occasionally be a little too involved in their immediate work to think much about its consequences. None of us like to think that the work that we are involved in may hurt people, disrupt lives or cause misery to our fellow man, and so there has perhaps been a tendency to play down the individual suffering that robot introduction may cause, while emphasising the long-term global benefits.

Such an attitude on the part of those who are in a position to truly appreciate the likely impact of robotics, does not seem strictly honest. There is little more objectionable than a small elite making decisions 'on behalf of everybody' without even publicising what the likely consequences are. There *are* occasions when, in practice, it is necessary, but the introduction of robotics is not one of them. It is highly questionable whether the doctor telling a child that the forthcoming injection 'won't hurt a bit', does anything other than breed mistrust. The same applies for robotics.

Whatever the global employment consequences, *of course* there are likely to be very large numbers of *individuals* put out of work either directly or indirectly as a result of robotics (although higher demand for cheaper products might reduce such unemployment). *That is one of the prime tasks for which robots are built – to replace humans.* Nevertheless, there is an irreducable element of conflict and difference of perceptions over the employment issues of robotisation, which depends (in its starkest form) on whether one is a manager or engineer charged with reducing production costs, or a worker faced with loss of one's livelihood. Yet for many there appears to be a lack of even appreciation of

this difference of perceptions, together with the underlying considerations of economics, creation and distribution of wealth, and politics, all
of which are inextricably tied up with the question, 'What about my
job?'.

Some believe that the replacement of humans by robots is not of
itself, such a bad thing. They stress that what must be remembered is
why replacing humans by robots is desirable in the first place, what the
long-term results would be if such replacement did not take place, and
why being prevented from 'working' should be considered so awful
anyway. To many who have studied these considerations, the inevitable
conclusion has been that large scale introduction of robotics is vital.
Nevertheless, even if they are correct, the eventual benefits of such
introductions would be conditional on many factors, not least political
ones, and so there can really be no guarantees as to the eventual
benefits. Additionally, there is certainly no implication that the achieve-
ment of those objectives will be painless to all concerned.

That is what this chapter is about – the hopes and fears of those
affected by robotics. Some of those fears seem to be justified, implying
that the 'Robotics Revolution' demands compassion, understanding and
above all realism to be exercised by its advocates throughout the
wholesale upheaval that it brings, and that true consideration must be
given to providing concrete relief to what casualties there may be. To
overdramatise the likely problems is irresponsible, causing unnecessary
worry and resentment. Yet to be complacent is to be unprepared, again
causing unnecessary suffering. Maybe the ideal should be for all who
aim at the global target not to lose track of the individual, while the
individual should not forget the common goal. Then again, maybe too
much idealism is out of place when discussing a man's livelihood . . .

So what are the specific fears of the individual? Over 30 years ago, in
1952, Kurt Vonnegut wrote a book called *Player Piano*. The story is set
in the future, and is a bleak vision of a world in which automation has
become so advanced that, apart from a small number of managers,
engineers, civil servants and professional people, the vast majority of
the human population lead lives without work, dignity or meaning. A
recent report by Ingersoll Engineers has commented in similar vein that
although unsocial hours and heavy work may eventually be substantially
reduced by robots, what takes its place could, in its own way, be even
less pleasant.

Yet such concerns tend to be long-term worries. What is of immediate
importance to most individuals is whether or not introduction of robots
is going to throw them into unemployment. Naturally, trade unions are
wary of new technology, and in the UK, Clive Jenkins pointed out that as

a union leader he believed that although the technology is very flexible, it is not necessarily neutral, and its benign nature cannot be taken for granted. Even in Japan, where for about a third of the working population the policy of 'a job for life' (popularised during the 1920s) has till now resulted in worker acceptance of robots, fears are mounting amongst certain car workers that robots on the assembly line are threats to job security. People who were at first delighted to have robots remove them from tedious or dangerous jobs, are now wondering if the robots will soon remove them altogether! Some of the smaller Japanese firms seem to be uncertain about introducing robots, and this may reflect the increased publicity given to the adverse social consequences which might arise from such introductions.

Such worries about people losing jobs cannot honestly be completely satisfied by assurances from managers that manning levels will only be reduced by means of **natural wastage**. Such agreements only shift the underlying problem from the company level to the national level – those who would otherwise have joined the firm to replace the workers who had retired now have no such vacancies to fill. *They* 'lose' the jobs that otherwise they would have got. The problem remains, but is far more anonymous. The possible implications of this for national productivity are discussed later in the chapter.

Why do it?

In deciding for oneself whether the problems which apparently 'necessarily' accompany widespread introduction of robots are nevertheless outweighed by the likely benefits, it is first important to remind oneself just what those benefits really are, what the incentives for robot introduction are, and why comparatively suddenly robots have become an attractive option. These topics have already been widely discussed in this book, and economic justifications are detailed in chapter 16, and so only a brief recap will be given here. It is also worth remembering, of course, that a rational decision concerning the above questions is likely to be very heavily influenced by just who is making the decision. Where the boundary between 'reasonable' and 'unreasonable' is drawn is not absolute, but must to an extent reflect the situation of the arbiter.

Perhaps one of the major incentives for introducing robots arose because the cost of employing a worker increased particularly steeply throughout the 1960s, especially in the USA and Western Europe. As a result, the cost of buying and running a robot gradually became a more attractive propostion to managers whose firms were struggling to remain competitive. A second reason, in some countries such as Japan, has been a steadily increasing shortage of available manpower. In Japan, the population is currently growing at about 1.1% per annum, but because so many more Japanese youngsters are going on to higher education, the workforce is in comparison only growing at a rate of 0.7%. Even those young workers that there are, today consider such jobs as welding and painting as 'dirty work' which they do not wish to perform. Workers with higher levels of education are even reluctant to study shop-floor work procedures, feeling that their educational standard should preclude them from such mundane activities.

Even in countries such as the UK, improved mass education has, rightly, led workers to want more out of life than to act as 'glorified machines'. It may seem ironic (or even infuriating) that during a period of high unemployment there should be manpower shortages in places, but the fact that there are large numbers of people without a job does not imply that those same people are willing to fill particular vacancies. To many it is understandable that school leavers and others should sometimes prefer to remain 'on the dole', rather than accept a job which they know is below their capabilities. It may frequently not be laziness,

Figure 13.1 Is sweating over manual arc welding just as degrading as being out of work?

but instead a pride in themselves. To some it may seem more degrading to fill one's lungs with paint in a paint-shop, monotonously assemble components like a machine, or sweat over identical welds in a metal-shop, than it is to be one of the unemployed. Being without a job can, of course, be very degrading, as will be considered later, but that is due to factors such as lack of a social group, little money and the attitudes of others.

For some employers, then, it appears that the robot has arrived just in time to fill those jobs humans do not want. Yet in many countries, labour shortage is primarily in skilled manual workers, suggesting that, far from being caused by a refusal of workers to take the jobs, it is possibly caused by a reluctance on the part of employers to train the necessary personnel (so in effect *creating* the shortage). Firms might, of course, counter this argument by pointing out that because robots can reduce the amount of unskilled work performed by craftsmen, these men can be freed to carry out more tasks requiring their full abilities.

It has been suggested that a third reason for increased robot introduction may be the growth of legislation and awareness about health and safety in factories. Robots can take over from workers who would otherwise be exposed to toxic fumes, excessive heat or physical danger. Similarly, there are demands, in some countries at least, for greater job satisfaction. Robots can quite happily be used to carry out the most tedious of tasks without becoming alienated, staying off on Monday mornings or just becoming bored stiff! Nevertheless, various studies, such as at Volkswagen in Germany and by Carnegie-Mellon University, USA, have raised doubts about the capacity of robot introductions to improve job satisfaction of the human workforce that remain. It was found that in practice mental stress was often increased. Such considerations about the possible effects of robots on the human workforce are dealt with later in this chapter.

Social implications

Japan is the world's leading user of robots; it also has one of the lowest levels of unemployment. Nevertheless, as far back as 1981, an editorial in the *Japan Economic Journal* stated, 'For the time being, an over-whelming majority of people seem to be supporting the introduction of robots, saying the immediate impact of robot introduction upon em-ployment is very light . . . But despite this favourable position, a prob-lem remains as to whether this favourable environment for introducing robots will continue forever. It depends on the pace of overall economic growth in the future, but it is safe to say that the Japanese economy will steadily lose room to absorb workers displaced by robots in the long run'.

At present the introduction of robots has been sufficiently small (even in Japan) that the effects on unemployment and other social aspects have been minimal. Clearly, this will not remain the case for long. The United Auto Workers in the USA reported that the introduction of 85 paint-spraying robots into 18 General Motors plants resulted in the disappearance of 211 jobs (almost 12 workers per robot)! Force Ouvrière in France have estimated that one robot removed about 3 jobs, and the Finnish metalworker's union placed the figure at 3–5 jobs. A US study in Michigan has predicted that by 1990, between 100,000 to 200,000 manufacturing jobs will disappear in the USA, mainly in car manufacture – around 40% of the painting jobs and 15–20% of welding jobs will be extinct.

In Michigan, where the recession has caused a major slump in the car industry, much effort is being directed towards turning the state into the robot manufacturing centre of the USA. A state survey indicates that, even once all the new jobs created by the project are taken into account, if plans are successful then 50% more jobs will have been lost than will be created. A further study, supervised by Donald Smith, Director of the Industrial Development Division at Michigan University, suggests that the number of workers directly displaced by robots by 1990 will be 7%, but 88% of these will be redeployed within their present company, 63% of them working with robots.

A study by Commerzbank, West Germany, claims that about half the 1.2 million assembly-line jobs in Germany will be threatened before 1990, with each second-generation robot replacing from five to ten workers each, and taking over 90% of spot-welding and paint-spraying jobs. Meanwhile, Volkswagen estimates that second-generation robots will do 60% of all work in the car industry, while an OECD study predicts that even by 1986 at least 1.5% of Japanese production workers will have been affected by robots, by 1987 15% of assembly jobs across all US industry will be performed by robot, and by 1990 200,000 jobs will have been lost because of robots in West Germany alone.

These predictions have not, of course, been tested, and so the apparent underlying trend may in fact be spurious. Nevertheless, as they stand, they do not hold much comfort for the person already out of work. There is not a great deal of hope that once robotised factories start expanding they will take on their old workforce again. Recession may have hastened demanning, but many believe that the recovery involves avoiding massive intakes of labour when demand rises, and using robots instead. When recession is prevalent and unions conse-quently suffer a loss of bargaining power, there is also the possibility of managements forcing through technological changes without consulta-tion, reducing union membership still further.

A new revolution?

In the UK in 1811, a period of great distress, a secret association of working people was formed with the object of destroying the new textile machinery, then being largely adopted, which they regarded as the cause of their troubles. The first outbreak was at Nottingham and was inspired by the action of a supposedly mentally retarded textile worker named Ned Ludd who, 30 years before, had smashed his machine to pieces, so giving the name 'Luddites' to the movement. Afterwards, serious Luddite riots occurred in various parts of the country, especially Yorkshire, where many people were killed, mills destroyed and several rioters tried and executed. One opportunist, a blacksmith named Enoch Taylor, both made machinery for the mills and also the large hammers which the rioters used to destroy them! Hence the Luddites' cry, 'Enoch made them – Enoch shall break them'. Yet if the Industrial Revolution, which started in England around 1760, could result in such a drastic counter-revolt, is the 'Robotics Revolution' really comparable?

The Industrial Revolution had a truly incredible impact on the way of life of those caught up in it. Within only 90 years, by 1850, the UK changed from an agricultural country with nearly 70% of its people living in the countryside, to an industrial land with only 10% of its population (which had by now doubled to 18 million) working on the land, and over half living in the new towns and cities. The number of towns with populations of over 50,000 increased over the same period from two (London and Edinburgh) to 29. The great manufacturing towns sprang up in the North, with Leeds based on wool, Manchester on cotton and Birmingham on engineering. Liverpool became a major port. Britain produced 70% of the world's coal and cotton cloth, 50% of its steel, and 40% by value of its hardware.

Yet a new spirit of inhumanity arose, driven by 'economic necessity'. Relationships between labourer and employer became harsher and more impersonal. Although there were slight improvements in average material standards, these were more than countered by increased ex-ploitation and oppression, and craft workers found themselves operat-ing machinery which both took the joy out of their work and even threatened their livelihood. As the introduction of steampower brought about still higher output, increasing numbers of craftsmen were displaced. In many cases it was a burning feeling of injustice about intolerable conditions, rather than an irrational or short-sighted resistance to progress, which drove such starving people to revolt and attempt to slow the technological advance.

Yet by the 1840s nearly 50% of the UK workforce was in **secondary industries** (manufacturing, construction or mining). Almost a century and a half later only about 40% of the working population in the UK are in secondary industries, a few percent remain in the **primary industries** of fishing and agriculture, yet almost 60% work in **service industries** (sometimes called **tertiary industries**) ranging from offices to shops. Will this gradual trend away from factories and into service industries be dramatically accelerated by robotics, so that the effects of this 'revolution' are as marked as during the original industrial revolution?

A report on new technology by the Council for Science and Society (CSS), UK, distinguished three major kinds of reply. Some people believe that the magnitude and speed of the changes have been exaggerated. The move to the service industries is not a 'revolution' but is instead a gradual evolutionary progress – one more stage in the development of industry, comparable to other developments which have gone before. This argument is backed up by reference to the widespread expectation in the 1950s that automation would bring about rapid and fundamental changes in society – which it has yet to do.

The report's second (opposing) stand is that we are indeed moving out of the stage of the industrial society and into a post-industrial stage. The supporters of this viewpoint believe that in the future we will only need approximately, the same size of workforce to produce goods as we currently use to produce food (about 5% of the whole). The remaining 90% will all work in the service industries, many of them in some way involved with the storage, manipulation and dissemination of information. In such a society, it is claimed, the two functions of work (production and earning) will probably be divorced, with leisure becoming widespread and work being only a minor and occasional diversion.

The CSS report's third school of thought agrees that we are in the midst of a revolution, but predicts a total breakdown of society. They claim that massive unemployment will cause unrest, leading to the breakdown of government and the emergence of some form of dictatorship from which, through a revolution, will arise the kind of post-industrial society mentioned before. Clearly, none of the three above viewpoints is testable. In reality, it seems likely that the actual course of events will lie somewhere within the three extremes, but it is truly difficult to discount any of the viewpoints as being completely impossible. Since robots were first introduced, for instance, people have wondered what reactions they would evoke among workers, and worried whether a **neo-Luddite** reaction would ensue.

In some cases it has. Robot numbers are still low, so the reactions have similarly tended to be small, and it must be said that much evidence is only anecdotal. Some reaction nevertheless has occurred. Some notable examples have been in Eastern Europe, where attempts to thrust the bright schemes of bureaucrats onto unconsulted industry has increasingly resulted in sabotage. In one Ukrainian engineering plant, a routine order for a press resulted in the arrival of a brand new press complete with robot to service it. The factory had to take what it was given, but quickly uncoupled the robot, worked the press manually, and threw the robot on the scrap heap. Elsewhere robots mysteriously break down, and nobody bothers to send for spares.

Yet in some of the early introductions of robots in the USA as well, such as at the Ford Motor Company plants, workers initially physically sabotaged the robots, until they learned to understand them. Two years later the workforce was pinning 'Get Well' cards on to the robots when they broke down! However, in the early days, Ford felt that the term

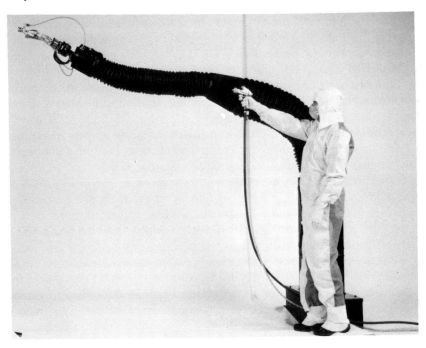

Figure 13.2 Man replaced by machine?

H

'robot' was so sensitive that they coined the alternative term of 'universal transfer device' to use in discussions with their workers.

Not suprisingly, unions throughout the world are taking the question of automation very seriously, yet are not really opposing it. West Germany, for instance, has reported that 53% of its workers are afraid of the effects of new technologies on their jobs, yet its metal-workers' union has stated that, 'There should be a shift from a defensive strategy against rationalisation to an active strategy to deal with innovation'. The International Metalworkers' Federation (IMF) has said, 'There are no IMF unions which resist totally the introduction of new technology as such, but most of them watch closely the development and the introduction of new technologies. . . .' One of their worries is that after robot introduction, although they may manage to keep their jobs, they may be left with tasks which are even more boring than before the robotisation. This problem is tackled in the next section.

A history of deskilling

Division of labour is an extremely ancient practice, stretching right back to when men hunted and women collected fruit and cared for children. During the industrial revolution, this process was extended to the degree detailed by Adam Smith in 1776 in a well known description of pin making; ten separate people were employed in the manufacture of every pin, each person only responsible for a minute subdivision of the work. Such subdivision of work makes economic sense to an employer. Imagine an arrangement where ten people are each performing a task which is largely unskilled but requires 10% of the time spent on skilled operations. If each person performs the whole task, then they must each be capable of skilled work and payed accordingly (say two units per hour). If the employer now subdivides the work so that nine unskilled workers (paid at only one unit per hour) perform most of the tasks but leave all the skilled work to just one skilled worker, then there is a saving of nine units per hour. It is worth noting that if payment to the group were based on work accomplished, then it would remain (in total) the same for both cases. The advantage to the employer is not due to any improvement in technical efficiency.

This overall philosophy, which of course greatly deskills the operations required of the average worker, reached its limit with the production lines of Henry Ford. On these, a given worker often performed little more than an individual motion: 'The man who puts in a bolt does not put on the nut; the man who puts on the nut does not tighten it'.

Even present-day factories frequently include highly deskilled jobs, although some unions have made moves to prevent this, such as IG Metall in West Germany, who have an agreement that no human cycle times less than 90 seconds are permitted. Nevertheless, those deskilled jobs which do exist people argue, are ideal for robots. Many feel that with the advent of the economic robot, we at last have an opportunity to release humans from the kinds of jobs which they should never have been expected to do. Extending this argument they urge that, as subdivision of work results in jobs unsuitable for humans, we should accelerate the introduction of robots towards the ultimate goal of automation without the need for any workers at all. Thus, the impact of robots on the workforce will be entirely beneficial.

There are, however, counter arguments. Firstly, robots actually eliminate people from employment (which will be considered later). Secondly, it is claimed that it is not at all clear that the majority of jobs created around a robot are any more skilled than the larger number of jobs it replaced. Workers are often expected simply to bring parts to the robots and then take them away again. Thus, whereas deskilled jobs are indeed being taken over by robots, it is claimed that new such jobs are created as well, with workers feeling more isolated, more machine-paced, and more in fear of losing their jobs.

In addition, because for some workers skills based on analytic or logical ability rather than experience are increasingly required, there is always the danger that such a situation can lead to a polarisation between highly-skilled jobs and low-skilled work. This is, of course, a problem of all the new automation technologies, such as office automation, not just robotics. CAD systems, for instance, reduce the proportion of junior draftsmen to senior draftsmen (who actually use the systems). Such a technique allows far greater control of a project by excluding several people who would otherwise be needed from the process – which is fine for all but those excluded! As second-generation robotic systems become increasingly prevalent, so, of course, the impact of robots in areas such as assembly may have still more dramatic effects on employment. And that is only the second generation . . .

Returning to the problems of deskilling, the question must be answered. 'Does it matter?' Many would argue strongly that it does. In 1979 it was reported in the *American Machinist* that in a machine shop in Lincoln, Nebraska, 'Mike Bayless, 28 years old with a maximum intelligence level of a 12-year old, has become the company's NC-machining-centre operator because his limitations afford him the level of patience and persistence to carefully watch his machine and the work it produces'. Although this is one isolated example, it does stress the level of deskilling which is often present. Yet there is increasing evidence that such jobs are both harmful to the individual, and so, indirectly, may be to the company. In a discussion of employment and unemployment, Marie Jahoda states, 'Studies of the consequences of low-level unskilled jobs indicate that many workers in such jobs constitute a stratum of society consisting of degraded, frustrated, unhappy, psychologically unhealthy people. Their lack of commitment in employment colours their total life experience and they are unable to provide a constructive environment for their families.'

If such is the case, then not only is the deskilling a waste of human ability, with consequent losses to industry through poor workmanship, lack of motivation, absenteeism and labour turnover, but the boredom and alienation resulting from the trivialisation of work damages the worker and thus society. The css report mentioned above poses the question 'If it is unseemly for a civilisation to be founded on slavery, is it not also unseemly for a civilisation to be founded on work which is so far below the abilities of those who perform it?'

Economic determinism

There appears to be a 'roboticist's dilemma': either continue to robotise in the way that has already begun and so probably cause deskilling and unemployment, or else allow industry to continue largely as before, with the result that productivity increasingly suffers until such time as whole businesses crash, so eventually causing still greater unemployment. If the dilemma is truly a statement of the facts then the only option appears indeed to be to robotise; to do otherwise is to court still greater hardship and misery than may occur through the introduction of robots.

To many this 'economic determinism' suggests that, if we are to compete internationally, then we truly cannot afford the luxury of more desirable working conditions. Indeed, they say, it is necessary to intensify still further the elimination of human skill by means of robotisation in the interests of greater efficiency. We are constrained by economic forces. The current situation has arisen in response to competition, and consequently constitutes the most effective approach. It is necessary and inevitable, and any proposal that an alternative approach should be employed which made better use of human resources would entail an economic disadvantage.

Such an argument recurs again and again in reference to the possible unfortunate 'side effects' of new technology. Thus, back in 1956, Paul Einzig writing about automatic factories said, 'Automation is bound to

Figure 13.3 Is total automation the only economic choice?

proceed sooner or later whether we like it or not. Our choice does not rest between automation and full employment, but between prompt automation with the possibility of moderate temporary unemployment, and delayed automation with the certainty of grave perennial unemployment until our progress has caught up with that of our competitors'.

Over a quarter of a century later these words were echoed in a key passage of an ACARD report advising the UK government to promote robotisation actively, which read, 'It is sometimes argued that the United Kingdom faces, as a nation, a genuine social choice – either to adopt new labour saving technology and become a high productivity economy, but with the associated possibility of increased unemployment, or to reject such technology in favour of more labour intensive production methods, with lower real wages. We do not believe that in reality the choice is between two acceptable alternatives. We believe that the greatest threat to industrial employment in the United Kingdom arises not from the displacement of workers by machines, but from a further loss of world market share for our goods consequent on our failure to compete on price and quality. A key factor in reversing the decline is the adoption of modern production methods'.

Such appeals to economic determinism appear to be incontrovertible arguments of 'cold realism', yet after detailed consideration some people have begun to question their validity. Firstly there is the confusion of concepts all bearing the heading of 'productivity'. One interpretation of the term stresses the benefit to mankind brought about by a reduction in effort required for essential work, so freeing time for more desirable activities. A second usage refers to increased productivity of a particular organisation resulting in a more competitive position. This is not the same concept as the first, because faster 'pacing' of a workforce or subdivision of jobs will both bring about apparent increases in productivity, although in one case effort has been increased, while in the other wages are reduced although overall effort has remained the same.

A third usage of the term 'productivity' is in the context of a nation as a whole, in an analogous way to the individual organisation above. In this context, the country can increase productivity by producing more goods but with the same number employed and at the same wages. If on the other hand, the *same* amount of goods is produced by using fewer workers, the analogy breaks down for, unless they are 'guest workers' who can be sent home, the displaced people must remain in the country and so require support. The country's competitive advantage is therefore limited to, and indeed is based upon, the drop in living standards of those displaced. Thus, a country (as opposed to a company) can only effectively employ robots to produce *more* goods than before, and so try to obtain a larger proportion of the world market – a goal which is clearly impossible for all nations to succeed in reaching.

The sometimes conflicting interpretations of 'productivity' may result in success within an organisation creating disadvantages from the viewpoint of both society's interpretation of the term, and also of the national economic position. Thus, even at this level, some argue, economic determinism seems flawed, because the criteria used for 'economic loss' are too narrow; usually simply of an organisation itself. What would be more appropriate, they claim, is to maximise the joint benefits of organisation, society and nation. This would require a new kind of technology, avoiding the self-defeating lack of regard for human factors, which leads to stifling of initiative which could otherwise contribute to still higher productivity than the technology on its own can offer.

Yet, even without the problems of conflicting interpretations of productivity dependent upon one's situation, there remain serious problems with economic determinism. It may seem obvious that because competition enforces the adoption of the most profitable technology and work organisation, any changes (for instance to improve working conditions) must, of necessity, impair the competitive position of the organisation or nation adopting them. Yet, in the longer term, this may not be true. Professor Howard Rosenbrock, of the University of Manchester Institute of Science and Technology (UMIST), has pointed out that, 'around the beginning of the century, the piston engine established itself as the most successful prime mover in the motor car. More recently, serious efforts have been devoted to the rotary engine, but this has to compete with the piston engine as it has developed through decades of engineering effort. Even in current terms, the development of the rotary engine has to compete with the development of the piston engine, on which many millions of pounds are being spent thoughout the world. The previous success of the piston engine, and its established position, prejudice the development of the rotary engine even if this is a potentially better device'.

In the light of such arguments, it becomes clear that technology at any time may be far from the economic optimum that *could have been reached* if earlier shorter term economic decisions had been different. If a series of short-term optimal choices are made, each enforced by short-term market considerations, it is quite possible for the long-term situation to be far worse that it might have been. Yet, because there is no other route to make a comparison against, at any one time it can appear that, because the current state is an improvement on earlier states, the series of short-term optimisations have irrevocably led to the optimum long-term result. In practice, a different ('nonoptimal') decision earlier might have led to a still more desirable final result. The implication is that now may be the time to make such a decision in robotics . . .

The wrong road?

Let us consider the 'roboticist's dilemma' again. In analysing all dilemmas for the correct course of action prior to making a decision, it is necessary to demonstrate that a dilemma truly exists – that is, there is no third alternative course. The arguments against the economic determinism of the 'roboticist's dilemma' seem to indicate that there is indeed a 'third option', which its advocates would claim was far preferable to 'the devil or the deep blue sea'!

Various suggestions for alternative approaches have been made, but basically there only appear to be two fundamental design strategies that can be adopted for automation: either technology is created to *replace* a person, or it is designed to *enhance* him. Up till now, technology has tended to be used to replace a human as soon as it is economically feasible to do so. While such replacement is from unpleasant tasks, such as is the case with many current robot applications, then there is relatively little disagreement over following the philosophy. As techniques become increasingly available for robots to take over jobs which are not so unpleasant, then, as discussed above, the issue becomes far less clear. Especially when, in addition, some robot applications do not totally replace the human, but merely relegate him to a position of coping with those few situations which the robot itself cannot manage.

It is relatively easy to think how to replace man; it is suprisingly difficult to try to enhance him. Nevertheless, an example of such work is that being conducted in the UK by Professor Howard Rosenbrock. His team is attempting to develop a **flexible manufacturing system** (FMS) in which the machine is once more subordinate to its operator. A simple **cell** has been constructed, consisting of a **numerically controlled** lathe, milling machine and robot. Instead of programming the NC equipment using a specialised computer programmer or a suitable computer-aided design system, it is instead programmed by monitoring the skills of an operator as he makes the first in a batch of parts.

This approach, in fact, takes far less time than the other two before production can commence, and does not require verification, amendment and reverification. In addition to approximately halving the usual programming time, higher machine settings (which with remote prog-

Figure 13.4 A conventional form of FMS cell – not the only approach?

ramming are necessarily conservative) allow significantly faster actual production times. Once the first part of a batch has been made, the operator is then free to use his skills in scheduling production. In this way it is felt that the precious resource of human skills and abilities is being better used.

Naturally, some might counter that if a computer can, with sufficient sensory input, schedule a factory better than a human (which it possibly already can) then surely it is ridiculous for a human to insist on trying to perform the task himself. If an FMS, in due course, will in fact be capable of being programmed faster and better than an experienced craftsman could do it (which some feel is certain) then is it not really a forlorn hope to try to maintain the status of the human?

To an extent, though, this misses the point that people like Rosenbrock are making. Certainly, they would say, a computer may be able to schedule better than a human alone can, but that is not to say that a system could not be devised comprising both a computer *and* a human which was better still! There is a direct analogy here with computer systems designed to aid doctors in diagnosis – no one suggests that the doctor is no longer required. Indeed, although one can envisage such a diagnosis system which had been so optimised that it could be operated by a mere technician, without doctors there would, of course, be severe problems in extending medical knowledge.

Other options

Although many people believe that the current concept of 'work' will remain meaningful into the next century, there is also a strongly held belief that work as we understand it will have largely disappeared within 20 years. The argument is that because of the increased productivity and quality engendered by robotic systems, those that work with them can work shorter hours for the same pay. Indeed, in the UK, Terry Duffy, a union president, has stated that he believes a target of a 10% reduction every five years would not be unreasonable. This would mean a 26 hour week by the year 2,000.

Over the last 100 years, in many countries real wages have quadrupled while population and life expectancy has doubled, yet the average proportion of people's waking hours which they devote to paid work has been reduced from about 40% to under 20%. It is suggested that it will soon be possible to increase significantly the number of jobs in robotised factories by 'doubling up' what workforce there is, each group working perhaps only three days a week yet still taking home a good wage.

Additional jobs can be created in a similar fashion by introducing longer holidays, earlier retirement and increased further education, resulting in a shorter total working life. The extra wealth earned by all these people (thanks to robotisation) needs to be spent somewhere, so large numbers of extra jobs will be created in fulfilling the increased demand for services, especially in the leisure industries.

Some suggest that perhaps a still more flexible approach is appropriate, rather than automatically assuming that if at all possible everybody must have a 'full-time' job. Thus, although a task force of the Australian Labor Party has noted that work appears to be economically, socially and psychologically necessary for most people, and unemployment tends to be dreaded as humiliating and intolerable, it continues, 'We should abandon the masochistic doctrine of work for work's sake. There is nothing life-enhancing in performing boring and exhausting work year after year'.

Several people believe that we are heading for a multiple-activity society in which people's lifestyles may not principally be determined by the job they do. Many people may still wish to opt for paid work, where there will be the option of working longer hours for higher pay and then paying others for services which one does not have time to do oneself. There is also the other option of working shorter hours with the flexibility of lifestyle which that affords – for instance indulging in do-it-yourself rather than paying others to do it for you.

Much more time, it is claimed, will be spent learning – not necessarily 'academic' work, but also new training for one's job. Technology is changing so rapidly that the idea of being trained only once at the beginning of one's career is already largely redundant. Indeed, many people may in fact have two or more careers throughout their working lives, and many more are likely to be self-employed. With the increased availability of communication, it may well be that the majority of people will move out of the cities into smaller communities, and work close to (or even at) home.

Nevertheless, there are some problems with such scenarios. Although there have been cultures, such as Classical Greece, which did not value work highly, the traditions of modern society have tended to place work in a central position which many would find hard to dispense with. Employment tends to enforce a structure of time organisation – when this is removed its absence represents a considerable burden for most people. In addition, employment currently offers social contacts, a social environment, status and prestige, all of which would need to be replaced.

Of course, many professional and technical jobs require knowledge and practice which simply could not be obtained from a few hours work a week. People in these types of jobs would still need to work for long hours, and presumably benefit accordingly, maybe causing resentment from those for whom such jobs were not available. Allied to this is the problem of wealth distribution in general: if the link between work and wealth were completely broken, then the potential consequences are far from obvious.

The road to follow

In the USA, the United Auto Workers has negotiated retraining programmes, paid for by companies themselves, and open to all laid-off workers. Special deals have been made with such firms as Ford, International Harvester and Mack Trucks, while at General Motors a $10 million training scheme was arranged in conjunction with the State of California for 8,400 workers affected by closure. An unprecedented agreement has been made by the Nissan Motor Company in Japan, that no employee will be fired or laid-off due to the introduction of technological innovation, with guarantees against demotion or reduction in wages where job transfer is necessary.

An awareness is growing. Some of the potential problems of robotisation are being discussed and tentative solutions are being suggested. Mutually satisfactory safeguards such as those outlined above are being agreed between workers and employers, so that company goals can hopefully be achieved without individual hardship. Yet if there is one point that should be obvious from the preceding chapter: there is NO single obvious 'road to follow' towards the robotic future. More than ever there is the need for those in a position to influence that future, to try to bridge the gap between the different perceptions of it that exist.

14

Managing with robots
ROBOTICS AND THE FIRM

Success and failure

The slogan 'automate or liquidate' does not refer to two mutually exclusive possibilities! Why do some robot introductions fail, while the same application elsewhere is a success? Why can some firms bring in robots with the support of their workforce, while with others it is a battle all the way? Why were some countries (such as the UK) initially so much slower than others to incorporate robotics into industry? This chapter attempts to find solutions to these and other similar questions, but in one word the answer to each question is – 'management'.

Because widespread adoption of robotics is a recent phenomenon, there is still only a comparatively small pool of robotics-related management experience on which to draw, and in addition there have been extremely few specifically robotics-related management studies published. Nevertheless, it has become quite clear that the mere decision to adopt robots does not automatically result in a success. Indeed, when the Technology Policy Unit of the University of Aston, UK, conducted an analysis of the patterns of success and failure of 40 firms adopting or considering robots, they discovered that only 56% of the 32 companies which actually adopted robots met with straightforward success in using robots for actual production. In 44% of the cases, initial failure was experienced, and half of these abandoned robots altogether. For those that 'kept at it', eventually successful installations were usually far simpler than originally envisaged.

Clearly, many firms experience considerable difficulty in successfully introducing robots, and it is slowly becoming apparent just what types of problems they have in common. Technical difficulties seem to be experienced in both successful and unsuccessful introductions, and long development periods of up to two years are common. Lack of previous experience of automation, however, appears to be a great disadvantage to firms considering robotisation, and in many cases may lead to premature 'throwing in of the towel', when initial enthusiasm is not rewarded as quickly as would be liked.

83% of the successful firms in the Aston study already had automation, while only 25% of the unsuccessful firms had previously automated. Not only does existing automation imply the likelihood of opportunity for robots to service these machines, but previous experience is likely to cause managers and engineers to have a more accurate expectation of robotics, and it also implies the presence of appropriately competent people. As would be expected, considerable managerial effort is required during robotisation, although this returns to normal once the development phase is over. In other words, during robot introductions it is really the whole organisation which is in a learning period.

Although, as mentioned in the last chapter, worker resistance is not common, when it *is* present it frequently seems to contribute to failure of a robot project. In practice, especially when workers are provided with a commitment about job security, they frequently react with interest and curiosity. If anything, management appears to be overly concerned about the negative response it will receive, and this may in fact be a greater inhibiting factor than labour reaction itself. Nevertheless, there are other forms of equally (or more) harmful resistance, including organisational barriers (such as between R&D and Production), daily immediate company problems overriding those of the longer term robot programme, personal commitments to existing systems and, most importantly, lack of technical expertise amongst management. Some analysts have identified this last problem as a major failing in countries such as the UK, resulting in unrealistic demands being made for robot performance and installation.

Few firms bother to employ sophisticated or sensitive methods of evaluation either of the cost effectiveness of an introduction or subsequently of its performance. Nevertheless, when such methods have been used, the Aston study indicated that they almost always led to a successful project, and so lack of sufficient evaluation must be considered to be an unfavourable trait. (A detailed examination of different economic evaluation techniques is provided in chapter 16.) Another major difficulty is frequently the lack of suitably trained personnel with the required knowledge of mechanics, electronics and programming. Maintenance can therefore be a problem (despite special provisions for training), especially when, because of the shortage, carefully trained men are offered a better paid job elsewhere!

A fundamental problem, contributing to the sluggish response of large manufacturing companies is the structure of the companies themselves. Traditionally, such organisations are of course based on a multilevel hierarchy, with ultimate control resting with the chief executive, and passing through several chains of command to the shop floor. Although this structure enables top management to control a whole company, it is well known that the necessary distance between them and the lower levels of the hierarchy makes it very difficult to remain in close touch with the day-to-day problems of the shop floor – indeed this is often largely deliberate.

In addition, of course, communication between different branches of the hierarchy (for instance, different departments) can often be poor, and resulting friction can hinder the functioning of the organisation as a single unit. Thus, while a young and dynamic company (typically with few levels of command) may be able to respond rapidly to market forces and incorporate robotics quickly, a larger more 'vertically structured' organisation often finds it much more difficult to display the same degree of flexibility. Nevertheless, in the Aston study it was considered that the outstanding advantage of robotisation was the *enhancement* of managerial and technical control. Improvement in process control was found in 52% of the successful robot introductions, and improvements in quality and consistency were also apparent. These improvements were attributed to the elimination of human variability in the production process, also reducing dependence on human labour and vulnerability to possible strike action.

Nevertheless, it is worth noting that studies conducted at Case Western Reserve University, USA, pointed to the relative weakness of estimates of expected technological benefits of innovations in general. The most common sources of such errors included underestimating (often considerably) the time needed to achieve effective functioning of the innovation, overestimating the average utilisation rate and underestimating the requirements for adaptive adjustments in the preceding and subsequent operations of an integrated production process.

Are you ready for robots?

Increasingly, management is well aware that robots can improve productivity and efficiency, increase product quality, eliminate hazardous or unpleasant work and overcome labour shortage. However, although some pioneering installations can be justified simply by the experience in robotics which they provide, it goes without saying that robots should usually only be installed where they are cost effective. Selecting suitable applications, however, is not always obvious, and there are also times when, despite a desire to install a robot, a company really ought to 'hold off' for a while until it is more suited for such an introduction. Individual robots, for instance, are rarely economically viable, as they may require almost as much outlay in training, provision for maintenance and so on, as a group of robots would. In addition, they tend to provide *far fewer* 'hidden benefits' than several intergrated robots would (part of the **integration effect** discussed in chapter 16). In such cases it may be better for a company to wait until it can provide suitable applications for *more* than one robot.

For the currently available robots, suitable tasks will tend to be highly repetitive and labour intensive, and preferably will be unpleasant or

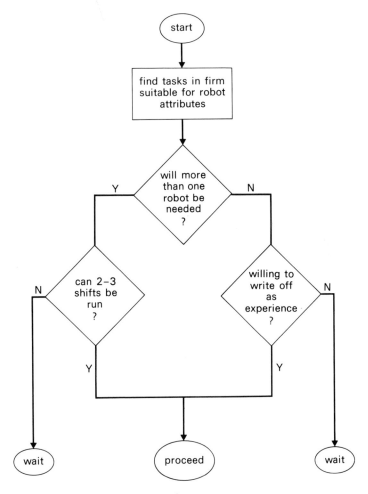

Figure 14.1 The decision to robotise

hazardous. Examples of many such applications have already been covered in detail in chapters 9 and 10. When looking at a particular application, the special attributes of current robots should be considered: their ability to handle heavy objects accurately (feeding a furnace?), follow a consistent path (paint spraying or arc welding?), place parts in precise positions (palletising?), be readily reprogrammed (small-batch production?) and so on. Potential tasks for a robot should (initially) be *simple*, require little modification to existing installations and minimal ancillary equipment. For the next couple of years at least, that may effectively rule out *complex* robotic assembly as a first application, unless a company can 'buy in' expertise.

It remains difficult to justify a robot which will only be used for single-shift work. Consequently, a company must assure itself that it can indeed run the robots it intends to purchase for at least two (ideally three) shifts a day. In reality, this rarely costs substantially more than running them for single-shift work, but there must, of course, be a market for the increased production – otherwise the whole exercise becomes meaningless! Having answered all the above points honestly (there is no point being carried away on a tide of enthusiasm) then at last the prospective company can decide if it really is quite ready for the introduction of robots. For a company to proceed unwisely is to leave itself open to disappointment in the future; the whole enterprise will probably turn out to be an expensive mistake, which may effectively hold up future robotisation of the company when it does at last become justified. On the other hand, to take the over-conservative route and not proceed when it would be reasonable to do so is to give one's competitors a distinct advantage.

Forward march

Just as each company is different, so is each robotic application. Nevertheless, it is of course possible to suggest a broad overall strategy which management should follow, as far as they consider applicable, when they are looking at the possibility of robotisation. The sequence of stages through which the management of robotics should pass tends to remain consistent for different applications, although various stages are likely to exhibit increased importance for different companies.

Introduction approach

This stage is *vital*, and it is vital that is comes *first*. The moment a company starts *considering the possibility* of robots is the time that representatives of everyone who might be affected by robotisation should be drawn into the discussion. It does not matter that management 'don't want to cause a fuss because we're only tentatively looking at the idea' – if there is the slightest possiblity of introducing robots (and of course there is, or why waste the time) then *now* is the time to start allaying fears. Nobody who has read the last chapter can be unaware of the understandable concern that the discovery of plans to introduce robots can cause. Concern leads to resistance. The way to avoid such resistance is to bring the workforce into one's confidence at a very early stage. If robots are not eventually introduced, then something has still been gained – the workforce are likely to appreciate the openness which has been afforded them.

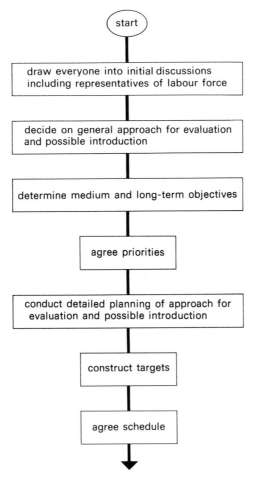

Figure 14.2 The procedure to robotise

Once everyone who may have something valuable to contribute to the discussion has been brought in (and all agree that there is no one else who should be included), then the general approach which will be adopted for the evaluation and possible introduction of robots must be decided. As with any major project, it is very important at this stage to try to ensure that there is strong consensus about the decision-making procedures which will be adopted, the budget, the evaluation schedule and so on. Once these complex issues (the details of which will naturally tend to be specific to a given firm) have been ironed out, then not only does everybody know where they stand, but the broad approach which is decided upon can, of course, act as a reference point against which actual progress can be compared.

It is interesting to note that the Japanese decision making process involves consensus by a large number of people participating in the discussions. Automatically, this results in a feeling of commitment to the decision by the group, so eliminating many possible objections during implementation. In addition, the large number of participants tends to prevent factors being overlooked, so generating an atmosphere of confidence in which major decisions can be implemented.

Long-term objectives

When considering introducing robots, a company must have a clear notion of what its medium- to long-term objectives for doing so are. Robotics cannot always be guaranteed to provide short-term benefits, so nothing more immediate than medium-term objectives should be considered. Such objectives may be concerned with productivity and efficiency brought about through higher throughput, better plant utilisation or reduced manpower requirements. Alternatively, robots may be seen as a way of overcoming labour shortages, of improving product quality or of cutting down on labour accidents or job dissatisfaction by eliminating hazardous or unpleasant working conditions. The main objective may simply be to gain experience in the use of robots. Whatever the aims, they must be explicitly agreed by all those levels of the organisation involved with the project. Otherwise, different groups might, naturally enough, have differing and possibly conflicting priorities for the robotisation.

Planning

At this point, in the light of long-term objectives and the agreed technique of introduction, it is worthwhile drawing up a detailed plan of how the evaluation and possible introduction processes will prodeed. In particular, labour savings and automation targets should be established, against which subsequent system selection and eventual testing can be evaluated. At the same time a detailed schedule should be drawn up, including not only target dates but also the selection of personnel who will be in charge of introducing the robotisation.

Understanding

It now becomes important to consolidate knowledge on robotics. This includes checking the different characteristics which robots possess (especially in comparison to the more familiar types of automation), the applications to which they are best suited and which of those applications are already 'tried and tested'. No book can ever be sufficiently up to the minute to guide the potential robot user on the latest available technology, so he must of course attend exhibitions, scan the robotics periodicals and obtain manufacturers literature direct. Although various useful publications have been compiled which contain copies of a wide selection of manufacturers literature, once again, these rapidly become outdated. In the list of further reading at the end of this book are a few useful sources of robotics information.

If possible, the potential robot user should also try to gain knowlege of the experiences of companies similar to his own which have already followed the robot trail. There are many valuable organisations designed to act as pools of knowledge for both new and old robot users – such as the British Robot Association and the Robot Institute of America. Joining such an organisation provides a direct link with the wealth of practical experience held by its membership, and also keeps one informed of potentially useful seminars, lectures and exhibitions. It should also be remembered that even though a direct competitor may not be particularly willing to give up the benefit of his experience, a lot can often be gleaned merely from what material the company has in fact chosen to publish – and what is conspicuous by its absence

Applications

Detailed consideration can now be given to the most suitable applications in the company for robotisation. Clearly, to have done so before a comprehensive understanding of robotics had been built up would have been to risk choosing inappropriately. It must be stressed that this is true even if a consultancy firm is called in. Such a firm may be an independent consultancy, but increasingly, industrial companies which have already built up robotics expertise through installing robot systems for their on use, are making their experience available through consultancy. In addition, many of the larger robot vendors will do a survey free of charge. Of course, the vendors will be biased towards their own equipment, but in practice, even the 'objective' consultancies are likely to favour those robots with which they have had more experience.

Nevertheless, no one should know more about the detailed operations of a particular factory process than the company owning the factory, so it is to the company's advantage to make sure that no relevant details are missed during selection of applications by the consultants. Secondly, there is the point that the consultancy firm *will not always be there*, so it is as well to start learning about robotics as soon as possible. There is an appropriate saying that, 'To catch some fish for a starving man is to feed him for a day; to teach him how to fish is to feed him for life . . .'

Potential applications should be viewed in the light of what opportunities they offer for manpower savings or for improvements in product quality, production methods, or work environment, and whether they

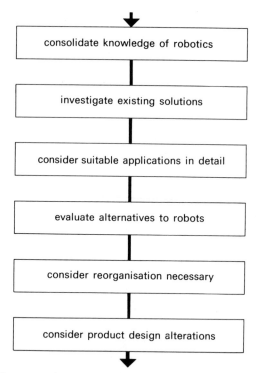

Figure 14.3 The procedure to robotise (cont.)

involve shift working. Alternatives to robots should also be considered, such as human operators, complex dedicated automation, or reusable (but nonrobotic) pick-and-place devices. For large scale robotisation, decisions should be made on which plant or plants should be automated, whether it will be necessary to build a new plant, or whether existing plants can be reorganised. In all cases, it will also be necessary to consider whether alterations to the designs of the products being manufactured in the selected applications could not make a substantial difference to the ease of robotisation.

Requirements

Having selected a small group of suitable applications, it becomes necessary to determine the requirements of each in terms of both a suitable robot (or robots) as well as any additional ancillary equipment and human labour. First, the physical requirements of the robotic equipment should be analysed, including the accuracy, repeatability, configuration, number of degrees of freedom, work envelope, speed, maximum payload, type of end-effector(s), flexibility, ease of programming, complexity of control system, memory size, sensory feedback, interfac-

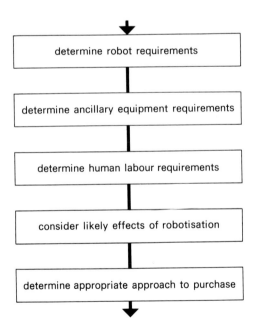

Figure 14.4 The procedure to robotise (cont.)

ing system, quietness, protection against unpleasant surroundings and any special requirements such as suitability for explosive environments.

Next, consideration should be given to the requirements for any other machines and equipment such as CNC machines, conveyors, jigs and other fixtures which are needed in addition to the robot. In some applications, such as robot feeding of existing machinery, the requirements for additional equipment will be very small; in others, such as robotic assembly, it may account for a major fraction of the total expenditure. Lastly, human requirements must be evaluated. A robot application may require maintenance personnel versed in mechanical, electronic, hydraulic and pneumatic skills, and a decision must be made on the level and extent of in-house technical support. The applications will also require programmers and operators, and, if the processes have not been completely automated, manual labour will still be needed to work alongside the robots.

Likely effects

Having obtained an accurate picture of what the potential robot systems would be like, it is worthwhile considering just what the implications of robotisation are. Demands for superior accuracy, reliability and continuity of earlier processes may be a problem, and special buffers may have to be provided. On the other hand, those same traits may be of great advantage further down the production process. To be suitable for the proposed robot system, products may need to be completely redesigned, standardised or simply made more consistent. In addition, use of a robot system for production may allow integration both with other production systems and also with **computer-aided design** systems, facilitating the eventual development of a full CADCAM system.

Of course, there are other effects of introducing robots. Many companies make use of their robotisation in their marketing strategy, hoping to portray a company image of progressiveness and high technology. Similarly, design changes and improved product reliability may substantially affect a firm's competitiveness and product marketability. Within a company, the introduction of robots is, of course, likely to result in transfers of manpower, and even redundancy. The Union response will largely depend on the methods by which these changes have been instigated – at both a local and a national level. Finally, on the shop floor, the multidisciplinary nature of robotics may cause problems in a workforce which has traditionally been heavily demarcated, as it may no longer be feasible to split tasks between different people in the usual way.

Suppliers

There are fundamentally three different approaches to purchasing an actual robotic system, which become increasingly distinct as the complexity of the systems increases. With the first method, the user purchases the whole system from one contractor. This necessitates only a small project team, but requires a great deal of faith to be placed in the vendor, with the additional disadvantage that there is a tendency for the user to abdicate responsibility at a time when it is vital that he remain more conversant than ever with what is going on.

At the other extreme is the traditional option of only buying in a number of small lots (robot, vision system, conveyors and so on), and then using a large in-house project team to dedicate their time to constructing the required system. Such an approach is becoming increasingly unpopular, because few companies carry the required extra staff, it can be very difficult to keep the project on schedule, managing a large-scale project using unfamiliar technology can be a formidable proposition and, at the end of the day, the user must take full responsibility for every system mistake that is made.

An increasingly popular compromise between the above two extremes is to set up a relatively small project team which specifies in detail the required robotic system, and then allocates a small number of self-contained contracts for different modules of the system, while remaining in control of the overall concept itself. In such cases it is naturally very important that vendors are selected carefully, as not only must they not underestimate the complexity of the task, but they must also adhere rigidly to the submitted specification (resisting the temptation to make alterations because it would 'make things easier') if their work is to integrate successfully with the modules supplied by other vendors. In practice, it is frequently better merely to submit a rigorous performance specification for potential vendors to quote against, rather than the user's idea of the best system to achieve those targets. This allows the vendors to apply ingenuity gained from experience in their particular field to possibly come up with novel solutions to the problems.

Cost analysis

It may only be once tenders have been submitted that it is possible to conduct a detailed analysis into the cost effectiveness of the prospective robotisation. This is such a major stage that in this book the whole of chapter 16 has been dedicated to it. The form of cost analysis employed will very much depend on whatever the company accountants consider

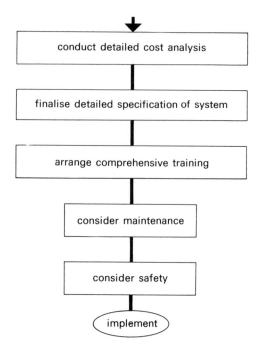

Figure 14.5 The procedure to robotise (cont.)

to be the most suitable methods of depreciation for the robots (with lives of probably well over a decade) and the remaining dedicated ancillary equipment (which may rapidly become obsolete).

In addition, it would be unreasonable if account were not taken of some of the indirect benefits of robotisation, such as the inherent flexibility of robots over their long working lives. In general, only top management will be truly able to suggest some of the wider benefits to company strategy of the proposals, and so, rather than merely having investment appraisals submitted to them, top managment should be actively involved in at least some aspects of the appraisals' construction. This implies the need for at least some advanced technological expertise at top management level. In the FMS *Report*, produced by Ingersoll Engineers, it states that 'it is no longer sufficient for the specification and justification for new equipment to be made at lower levels, leaving senior management to approve or reject it . . . senior management must participate in the analysis process and provide guidance'. It is notable that the large Japanese firms are making massive investments in appropriate education and training at *all* levels.

It is also worth noting that financial support from the government (in the form of grants) helps to reduce perceived economic risk. Indeed, in a study conducted by the University of Strathclyde, UK, it was found that some managers considered that **flexible manufacturing systems** would represent an uneconomic proposition without Government funding because of the high capital investment required.

Detailed planning

Problems concerning compatability between system modules are particularly prevalent in the areas of control and software – a vision system may work perfectly well on its own, but it is useless if it does not interface satisfactorily with the robot control program. Great care must be taken in supervising vendors as they proceed, lest they misinterpret their specifications. As any single vendor may be able to disrupt the whole programme, it is well worth the time and expense of trying to prevent such a situation by means of close supervision. This can result in greatly shortened eventual **commissioning** times. Still more time may be saved if all major equipment and software is tested and approved *at the vendor's site* before despatch to the user. It must be stressed that overall success seems to be largely determined by the quality and quantity of planning work, yet the amount of time required is nearly always substantially underestimated.

Training

As mentioned above, training and educating personnel in robotics, from top management downwards, is an increasingly vital, although extremely expensive, requirement for successful major robotisation, and as one manager has put it, 'If you think education is costly – try ignorance!' The Science Policy Research Unit, University of Sussex, UK, has stated that, 'Top management teams need to acquire the engineering skills necessary to interact with design and manufacturing management and engineers in order to develop coordinated automation strategies'.

Yet, at a lower level, simply securing numbers of skilled robot technicians may be something of a problem for many small and medium sized companies. Nevertheless, once the robot supplier is no longer supporting a system on a day-to-day basis, the company must, of necessity, find a supply of such people if it is to maintain its system at an optimal level. The solution is for the company to train its own personnel, and ideally plans for the training of technicians should be drawn up at the same time as plans for robotisation.

Robot training should be accomplished using in-house personnel if possible, but it is, in fact, frequently necessary to employ a robot supplier for the task, as he possesses unique information, experience and resources. In this case, it is most important that a mutually agreed document of detailed 'learning objectives' be drawn up. This agreement should specify in precise detail what is expected, in a given period, of both trainee and instructor. Trainees should then be monitored throughout training, using a prearranged set of performance targets, so that it is certain that, once a technician has successfully passed through the schedule, he is indeed up to the required standard.

Such an arrangement has been successfully employed by the Ford Motor Company at Dagenham, UK, who stress that it is mutually beneficial to both user and supplier, as, if there is a subsequent drop in performance of trained personnel, both know that training was sound, and so the robot user must look elsewhere for the cause. It is all too easy otherwise to blame lack of knowledge or skills when something goes wrong. The level of training required will, of course, depend on a worker's role. It is frequently impractical to rely on a maintenance contract with a robot supplier (if only because of the severe time penalty), and so many companies try to become self sufficient by training a complete in-house team of maintenance workers. Nevertheless, even when this level of training is not needed, it is a minimum requirement that every worker likely to come into contact with a robot should at least know how to turn it off or take other basic emergency measures. Safety can be substantially improved by sound training, and this, together with questions of maintenance, are among the points considered in the following chapter.

15

Danger – robot working/not working
SAFETY AND RELIABILITY

Safety problems

Unaware, unconcerned, the victim approached the welding robot. As he started to explain its operation to two fellow workers he could not sense the robot operator approaching the remote console, did not realise that no one had checked to see if there was anyone in the work area. Neither did the robot sense anything as it crushed the victim's chest and upper abdomen . . .

Elsewhere, a potential victim spotted a malfunction in one of the cutting machines serviced by a robot, and decided to correct the problem himself. Carefully he turned OFF the switch which linked the robot operation with the cutting machine, cautiously entered the danger area, and then quickly got down to work on the cutter, operating it manually. The robot lay impotent. At last the cutter was working properly. Another job well done. It should all work OK now. Is that everything? Flick the switch to link the robot and cutter again . . . Immediately the waiting robot crushed the victim between its arm and the cutting machine.

These two accidents actually occurred. The robots themselves, of course, were not the sinister preying machines portrayed, but the tragedies might not have arisen if the workers had thought that they were! One should cultivate a healthy respect for a machine which may be able to swing round a payload weighing a few hundred kilograms at a speed of over a metre per second. By the early 1980s only two people had actually been killed by being crushed by industrial robots, but a US jury ordered in 1983 that the family of a worker killed by an AGV should be payed $10,000,000 (£6,750,000) as compensation! The numbers of nonfatal robot accidents are far less clear, because frequently there is no special grouping under which such occurrences are logged.

Although robots are noted for their contributions to removing humans from dangerous workplaces, it must always be remembered that they are (currently) stupid though powerful allies. They may senselessly follow a sequence of instructions, not realising that they are wreaking havoc. Likewise, they may run amok due to actual malfunction of electronic or mechanical hardware or inadequacy of programming software. Of course, it is not merely the physical danger to human workers that is important here – robots also pose a threat to the products and equipment they are working with, as well as to themselves!

There are particular robot characteristics which can make them dangerous. For a start, a robot can move in an unpredictable path throughout a three-dimensional work envelope far larger than its own volume, unlike other machines, which usually operate in a predictable manner within an area surrounded by the machine itself. Robot movements can be so complex (especially if sensory feedback is employed) that sometimes even the robot operator cannot tell what the next movement will be. Particularly unpredictable movements occur when a robot moves from the end of a program back to the beginning again or onto a new program, or, following a power loss, moves back to its initial position.

A robot can appear 'dead', yet in reality simply be waiting for a particular sensor input; welding robots may suddenly strike an arc without the prewarning a human welder provides when flicking down his face shield. Chrysler's training programme dramatically emphasises the danger of thinking a dormant robot is safe: the instructor previously programmes a few minutes delay into the robot followed by major activity. During training he walks throughout the work envelope (keeping an eye on his watch!) and so lulls the trainees into a false sense of security. Soon after he has moved out of harm's way, the robot leaps into violent action – so making the point particularly strongly . . .

In addition to all the above safety problems, however much one attempts to separate humans and robots, there are occasions such as programming, maintenance, workpiece setting and tool replacement in which a human must enter a robot's work area, even though in some cases it is not feasible for the robot power to be turned off. Yet, of course, in addition to any moral obligation to protect one's workforce, different countries enforce various applicable legal statutes. In all of these, interpretation of the legal phrases themselves can be problematic, because such concepts as 'secure fencing' and 'danger' can be taken many different ways. In practice, the test of 'danger' frequently involves consideration of what is reasonably foreseeable in terms of human and machine behaviour. Again, it is usually impossible to remove totally any possibility of danger, but a 'reasonable' attempt should be made, balancing the potentiality and severity of accident against the time and expense of preventing it. A blend of safety with efficiency is required.

Safety solutions

In early robot installations, users rarely took account of robot-safety experience gained outside of their organisation. In addition, factory inspectors tended to be largely out of their depth, and in some cases insisted on excessively complex safety measures (so jeopardising economic justification), while in others demanded very little. With more practical experience available now, it is possible to suggest some of the approaches which may in fact be most suitable for minimising risk.

Enclosures

One of the most common approaches to robot safety is to employ an **interlocked enclosure** such as that illustrated in figure 15.1. Such systems surround the robot work area with fencing, through which there is an access door. The fencing is usually high enough that people cannot easily climb over it, and the door typically can only be opened once the power to the robot has been turned off, and the robot cannot be started again until the door has been relocked. If there is an operational procedure that whoever enters the work area takes the only existing door key with him, then, provided the procedure is observed, the robot cannot be accidentally started while someone is still in danger.

Figure 15.1 Example of an interlocked enclosure

Such a simple arrangement would probably have prevented both of the accidents mentioned at the start of this chapter. Nevertheless, various examples exist where the technique of putting into a cage what is considered dangerous (a practice little changed since the Middle Ages!) has proved ineffective. In one case, components left the robot work area on pallets, so a large opening was left in the fencing for them to pass through. In practice the operator regularly entered the work area via this opening, while the robot was running, in preference to using the interlocking door.

Programming

As mentioned before, however, it must not be forgotten that there are occasions when a human *must* work in close proximity to a moving robot. Programming is perhaps the most dangerous of these, and indeed a number of surveys have identified programming as the activity during which robot crushing is most likely to occur. The programmer's safety is then dependent on both the correct operation of the robot, and also the precautions that are observed.

While **teach-pendant** operation remains the most common form of programming, a major contribution to robot safety can arise from careful design of the layout of the controls. Careful application of ergonomic principles can reduce the number of operator errors, and also restrict the effects of those errors that do occur. It is important that an operator should never become confused about which direction the robot arm will move when a control is operated, whatever the position of the arm. Preferably the controls should all require constant finger pressure for continuous operation, so that without it the machine stops dead.

The maximum speed at which a robot can move during programming should be substantially lower than normal (possibly only 10%), and ideally this speed reduction should occur automatically. However, the correct teaching speed really depends on the robot application. When control of the robot is transferred from the main console to the teach-pendant, then it is important that the transfer should be complete, so that anyone at the console cannot affect the robot while a programmer is at work. Similarly, it is preferable that some form of interlock be used for control transfer (such as simultaneous depression of buttons on both console and pendant) so that accidental transfer cannot occur through knocking a switch.

Additional protection may be obtained by restricting the arc through which a robot can move, so preventing **overrun**. The best method of accomplishing this at present is to employ physical end-stops, but only as a last-resort restriction after other protections such as **software limits** (in which the controller constantly checks that the arm position is within specified bounds) have failed. In addition, it frequently makes sense in multiarm systems to arrange that only one arm at a time can be moved during programming.

Layout

Layout may dramatically affect risk. Although selection of a completely new layout is frequently not an option in many robot application because the robots are installed to tend machines which are already there, nevertheless it should always be given detailed consideration. The robot's maximum reach should not be too close to the perimeter fencing, walls or pillars, as an operator could become trapped or crushed – there should always be room for him to at worst be knocked away. Similarly, all pickup points for components (figure 15.2) should

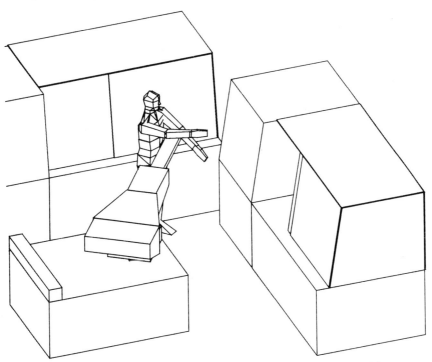

Figure 15.2 Computer generated graphics from a collision avoidance program developed at Nottingham University, UK

be thought of as potential trapping points. It may, in fact, be possible to arrange for a programmer to stand at the extreme range of the robot arm, so minimising the risk of accident.

As simple an act as puting up warning signs and marking the floor areas accessible by the robot, may enhance awareness of when an operator is potentially at risk, as will guardrails. A flashing light should indicate that a stationary robot is nevertheless active, for instance waiting for a prescribed set of conditions to occur. It is also possible to sound a horn at the start of each operation, but in practice workers soon tend to become accustomed to the noise and not register it.

Sensing

Increasingly sophisticated sensing can be used for safety purposes, although the adequacy of the systems must be of a high order. Because it is usually impossible to ensure that such systems are, in fact, **fail-safe**, care should be taken in their use. Although 'pressure mats', which stop the robot when anybody stands on them may be a useful additional precaution, they should not be used *in place of* a physical barrier. Similarly, ultrasonics or vision should probably only be used as support protection, specifically for those that have to work in close proximity to a robot, but not as a 'primary defence' against workers casually entering the robot work area. Various commercial sensing devices are becoming available for safety purposes, including a device which surrounds the robot work area with a 'curtain' of infrared light which, if broken, stops the robot. General Motors is developing an advanced system which detects changes in electrical capacitance caused by nearby objects, and so stops a robot if it comes within a few centimetres of a worker or other obstruction. However, such arrangements can usually only be used in relatively 'clean' environments – flying particles may otherwise trigger off the system.

Control

The control elements of the robot system should be made safer by means of employing only fully tested software, and reliable, high-quality hardware components. A robot can begin to behave erratically simply as a result of a power surge or dust inside the computer, so great care in the initial controller design should be taken to protect against environmental influences such as dust, heat, vibration and electrical 'noise'. Operation of a robot in a factory with a high ambient air temperature is likely to cause erratic robot operation, owing to the heat build up in the controller, unless appropriate cooling is used.

A particular problem arises when the control cabinet is open for servicing, so that the usual cooling system is rendered inadequate. Radio interference can also be a problem (especially with the cabinet open), so walkie-talkies and paging systems should not be operated near the robot. In addition, the microprocessor, memory and interfaces should ideally be constantly automatically monitored for malfunction, the whole electronic system should be adequately maintained and, most important, should be thoroughly protected from unauthorised use.

During robot movement, the controller should constantly monitor the arm for aberrant behaviour by detecting deviation from normal operation with regard to position and possibly also speed. In this way errors caused by incorrect electronic signalling, internal-state sensor malfunction or drive-system problems brought about by servo-valve, pump or power supply malfunction, are all likely to be trapped. Even simple point-to-point controllers can monitor movements by means of measuring the times normally taken to accomplish each move, and then comparing these with what is actually occurring.

Personnel

All workers who will come into contact with the robot installation must, in addition to their basic operational training, be made fully aware of the dangers of the job and of the suitable precautions they should take. There should be specified approaches to tackling fault conditions, rendering the system safe, maintenance, inspection and subsequent system restart. Workers should be trained not to bypass any safety devices (for instance, by climbing through an opening in the fencing rather than using the less convenient interlocking door). Similarly, there should be no unauthorised access to the robot, and what access there is should be determined by what is necessary rather than convenient. In addition, refresher courses are vital to remind personnel of procedures they had forgotten, and to train new personnel who missed the initial course. Because robots tend to require such infrequent major maintenance, it is not suprising that some of the safety procedures otherwise get forgotten between maintenance times.

Physical design

Robots themselves should not be designed with sharp edges, and any that are should have the edges padded. Other dangers include protruding linkages, exposed motors and transmissions, and loose cables and tubing. Grippers should be designed so that if there is a power failure the payload remains gripped – no-one wants a 100 kg ingot hurled at them in the dark! Similarly, care must be taken that even during sudden high acceleration a payload cannot slip out of the gripper jaws. Control consoles should be designed so that only those controls usually needed for robot operation are readily accessible; others should be concealed under transparent covers to prevent accidental use.

Finally, large red 'emergency stop' buttons should be located at the very least on both the console and teach-pendant, and they should be completely failsafe. It may well be that such a stop mechanism should not actually be a 'power off' switch, but a 'freeze' mechanism. In some cases, cutting all power to a robot could actually exacerbate an accident. For instance, a worker merely trapped by a robot might be crushed if total power loss caused the arm to sag under the weight of its payload.

Preferably, there should be some form of very obvious emergency-stop mechanism throughout the robot work envelope (perhaps like the 'communication cord' on some trains) which can be activated from any position, even if the teach-pendant is not at hand (such as that visible in figure 16.1). Just as important is that workers outside the robot cage can nevertheless deactivate the robot if they see a fellow worker in trouble. A tragedy of one of the fatal robot accidents was that the control system was such that nearby workers were unable to stop the robot from crushing their colleague.

Reliability

Despite initial misgivings, robots have demonstrated that they can indeed be reliable, even under hostile environments, and that even when they do malfunction, diagnostic routines allow the rapid determination of a fault and its subsequent correction. In designing robots for high reliability, there are problems caused by the fact that the environments in which even identical models of robot will have to operate can vary enormously. Nevertheless, over two decades various parameters have been determined which offer guidelines for robot design.

Many robots have to operate in environments which are potentially harmful to them, such as the alcohol–ammonia fumes present in investment casting applications, which can cause problems by attacking switch contacts, gear trains or bearings. Likewise, heat treatment often results in a hot, humid, salt laden atmosphere, making corrosion a problem. Spot welding causes molten metal particles to be thrown off, and some of these may bond to the robot, as may metal slag shot out from moulds during casting.

Another group of environmental problems are those made more hazardous because of the presence of the robot itself. These include tasks accomplished near naked flames (which could start a serious fire if a hydraulic robot leaked flammable servo-oil), as well as tasks such as painting (in which a spark caused by a robot could ignite the volatile paint solvent). Additional problems for reliable operation can arise due to shock and vibration (for instance from a hammer forge), electrical noise and interference (such as from electrical welding and starting of heavy machinery), and heat (from furnaces and the like).

To maintain high robot reliabilities under such conditions, various design strategies are generally adopted. Because of the high temperatures which may be encountered, servos and electrical devices are often eliminated from the ends of robot arms. Similarly, the robot control unit is frequently best located remotely from the arm itself, so avoiding excessive vibration, electrical noise and corrosive atmospheres, and it must be comprehensively protected against electrical 'spikes' on the power lines and 'pickup' on any wires linked with the arm.

When the atmosphere is heavily laden with particles, water cooling may be used in preference to air cooling, but all air should anyway be filtered, and the inside of systems maintained at a **positive pressure** to inhibit the invasion of dust particles. All exposed articulations should be surrounded in an envelope to protect them against abrasive dust, and hardened gears should be employed, as some dust is bound to work through. To protect against fire hazards, all robot covers should be nonflammable, as can be the fluids used for lubrication and hydraulics. Nevertheless, there is a substantial cost penalty in employing such non flammable fluids, which prevents them from being used as standard.

Actual reliability figures usually consist of the **mean time between failure** (MTBF) and the **mean time to repair** (MTTR). Both figures are important, as there is no point, for example, quoting a very long MTBF, if, when a fault does eventually occur, the MTTR is unacceptably long. In practice, industry is concerned with the proportion of potential running time that the robot is actually available (**uptime**) or not available (**downtime**). Experience indicates that a robot uptime of at least 97% is necessary to satisfy most customers. Thus, Unimation aimed at a 400-hour MTBF for their 2000 Series, assuming a downtime per incident (MTTR) of 8 hours (that is, 98% uptime).

Robot producers can estimate the reliabilities of their machines using techniques such as **fault tree analysis** (which is currently being investigated at Imperial College, London). By using data on the failure rates of individual components making up a robot system, together with an understanding of how various faults are interlinked, it is possible to

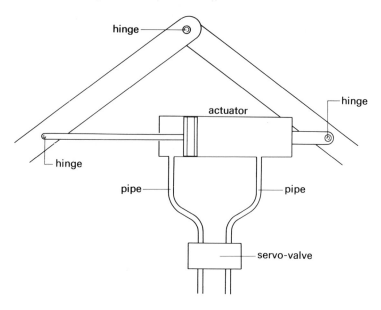

Figure 15.3 Possible sources of failure of a robot joint

determine the overall probability of failure of the whole robot. For instance, the stylised robot joint in figure 15.3 can fail either because of actuator failure or because of failure of one of the hinges. Hinge failure is a **basic fault** which can be assigned an independent probability. Actuator failure, on the other hand, can in turn be caused by pipe failure (which is a basic fault) or else either by structural failure of the actuator or valve malfunction. Although these last two can be traced back still further, it is in practice possible to assign them independent failure probabilities directly. Part of a typical fault tree for a robot (produced at Imperial College) is shown in figure 15.4. (The 'or' symbol indicates that the failure above it can be caused by any of the failures shown below it.)

Of course, faults can be software as well as hardware related, and increasingly the adequacy of 'supplied software' must be incorporated into any reliability analysis. As hardware complexity has increased, so too, fortunately, has its reliability, and so the increasingly complex robot controllers employed do not incur the unreliability penalties that might be expected. Nevertheless, although **fault avoidance** for both electronic and mechanical components (using highly reliable components and designs to minimise the probability that a fault will occur) is

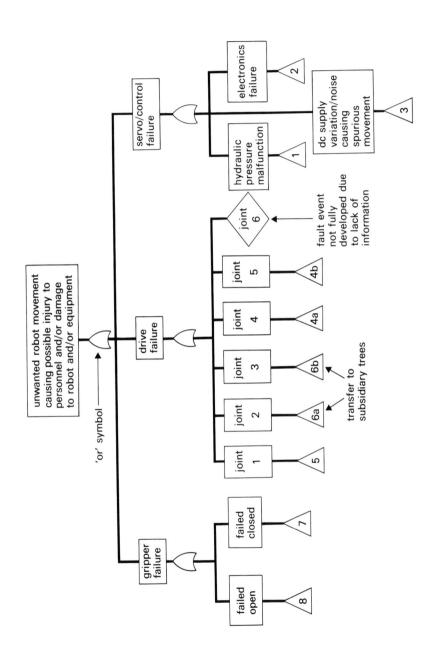

Figure 15.4 Example of part of a typical fault-tree for a robot

suitable for much robotics work, future applications such as those in space will require additional approaches, because no fault avoidance technique can totally prevent faults.

Such systems will need to exhibit **fault tolerance**, that is, to maintain system operation even in the presence of faults. Using sensors to monitor the robot's condition (including that of its primary sensors!) is paramount for such an approach. Fault tolerance involves fault detection, fault masking (to isolate errors to as small an area as possible), followed by fault recovery. Recovery is generally by means of **protective redundancy**, that is, switching out a failed component, subsystem, or entire system, and replacing it with a duplicate copy. Increasingly, of course, there is the alternative of using another robot to actually repair the faulty module. This is eventually likely to be far more practical (at least for the mechanical portions of the robot) than employing substantial redundancy.

Maintenance

Monitoring the MTBF of a given robot system may indicate the need for a major overhaul if a noticeable degradation of MTBF is discovered. In practice, such deterioration can occur at widely differing times, depending on the robot application, but an 'order of magnitude' figure is 10,000 hours, corresponding to about five man-shift years. Increasingly, however, the emphasis in robotic systems is to minimise the downtime of the overall system of both robots and peripheral machines, while maximising the number of production shifts. As a result, the maintenance strategies appropriate are becoming closer to those adopted for mass, rather than batch production.

There is a maintenance problem that arises when only a few robots are installed in a factory. Suppliers usually suggest that a trained maintenance man will not be effectively utilised unless he is responsible for at least three robots, and so for less than this number maintenance should be left entirely to the field service of the robot supplier. However, even for installations comprising more than three robots, they are comparatively so reliable that it is still difficult for a company to build up much maintenance experience, unless it uses a substantial number of robots.

A common suggestion is that robots should be maintained by a technician who can diagnose and correct faults in hydraulic, electronic, electrical, pneumatic and mechanical systems. Although automatic diagnostic facilities may aid him in this task, even the use of these facilities will require special skills. Thus the robot maintenance worker needs to be a specially trained multiskilled person. It seems likely that such an approach would lead to improved job satisfaction and motivation, but the introduction of such a role may be problematic. In countries such as the UK where the apprenticeship system still trains craft workers and technicians on-the-job, substantial reorganisation would be necessary, and many older workers might be unwilling or unable to expand on their single skill. Nevertheless, there are actual examples, for instance at Ford Dagenham, UK, where attempts at comprehensive training have hit fewer such snags than were expected.

A further question arises concerning centralised as opposed to decentralised maintenance. Usually, if the overall desire is to build up detailed knowledge about the workings and interelationships of a given group of robots and equipment, then it is preferable to employ a decentralised team of specialist maintenance workers who are concerned only with that group of machines. On the other hand, if the prime consideration is to reduce overall maintenance costs for the whole factory, then usually a highly centralised maintenance team is the optimum solution. Ford, USA, tend to install a robot temporarily in the maintenance department *before* putting it into its eventual production location, so that skilled maintenance staff can gain 'hands on' experience with the equipment. Increasingly, research is being conducted into employing computer simulation to evaluate the relative merits of different maintenance strategies.

Finally, the actual scheduling of maintenance is itself being reconsidered. There is growing evidence that regular routine maintenance can actually introduce reliability problems which would not have otherwise occurred. It is becoming progressively possible to employ automatic diagnostic and condition monitoring equipment to indicate when preplanned (but not strictly scheduled) maintenance should in fact be performed on a robot, and when, on the other hand, it is best to leave well alone . . .

16

Money makes the world go around
ECONOMIC JUSTIFICATION

Robots ARE different!

In one sense, this is the most important chapter in this book. However advanced the sophistication of a robotic system, however revolutionary its design, however progressive the firm for which it is destined, the sole purpose for which the robot was created is to perform tasks more economically than before. There is no point in a firm installing robots if, in the long term at least, the move does not result in an improved economic position. This is not to say, of course, that removal of humans from unpleasant or dangerous activities should not be considered as the most important justification (what value can be placed on a human life?), but, to be realistic, there are few firms which can afford the luxury of too heavy a commitment to worker satisfaction, unless this also either improves or at least does not lessen the company's profits.

It therefore becomes vital to understand some of the economic considerations involved in evaluating potential robotic applications. This is true not only for accountants, but especially for managers. A narrow and blinkered attitude may result in robots not being installed when they should be; maybe even worse, robots may be bought when an alternative yet less glamorous approach is, in fact, more cost effective. If mistakes like these are to be avoided, then those making the decision must appreciate just how much they can continue to apply their 'usual approach' to justification, and how much the radically different form of robotics technology warrants a reappraisal of these techniques.

Robots are reusable (figure 16.1). They are not like conventional automation, which is tied to a particular kind of job and becomes obsolescent after only a few years. Admittedly, in certain fields of rapid technological advance (such as robotic assembly), technical obsolescence of sections of the system may indeed set in after only a short time, but in many fields, such as machine tending or even paint spraying and spot welding, the 'older' robot may remain perfectly adequate for over a decade, simply being reprogrammed when new setups are required.

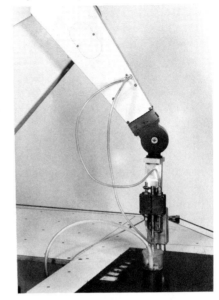

Figure 16.1 The same robot used for two radically different tasks
    (a) inserting rocker-arm assemblies
    (b) sensing and cleaning contamination

Clearly, approaches to economic evaluation which are not geared to take such factors into account may result in gross errors of judgement.

As customer demand forces increasingly rapid redesign rates for products (many product types such as televisions, which would continue to be bought for up to five years in the 1970s, in the 1980s require redesign after only two), so the advantages of a reusable robotic system become more attractive. Robots are not being used to fulfil the same requirements as fixed automation, instead they are allowing a flexibility which was previously unnecessary. In evaluating their suitabilities, it is not valid merely to compare the robots' economic performances if used with a firm's current-low variety product mix. Instead the comparison should be made between the cost of using either robots or the existing system to compete in the high-variety market of the near future. Such strategic (as opposed to tactical) considerations are of the utmost importance.

Many suitable robotic applications have in the past been incorrectly dismissed through over-simplistic evaluations, and an over-rigid adherence to 'company policy' for investment (which was compiled to deal with fixed automation and is hence largely inappropriate). It cannot be stressed too strongly just how unfortunate it is when those who are in a position to influence the adoption or otherwise of robot technology are unwilling to try to appreciate the different evaluation requirements of robot technology. Those, for instance, who dismiss an application simply because it does not fall within the 'payback period' traditionally used by their firm, may be doing themselves, and their company, a grave disservice.

In an effort to clarify some of these points, the remainder of this chapter is concerned with the various economic factors which should be considered when evaluating a robotic application, together with an explanation of some of the justification procedures commonly employed (in addition to why some of these are inappropriate!). Although the level of economics in this chapter is higher than in many robotics texts, it should truly be considered as the 'bare minimum' required by any conscientious manager in order to understand the implications of the evaluation procedure his company adopts. Discussions about techniques such as 'discounting' may at first glance appear very general, yet in practice these approaches are particularly appropriate for a *robotics* guide, because of their ability to cope with the robot's long useful life due to its unique reusability.

It is practically impossible to persuade companies to publish detailed figures of just how much a robotisation programme has cost them in total – in many cases this may be because the figures simply do not exist! More commonly, however, firms are understandably embarassed to disclose details of unsuccessful projects, yet wish to keep quiet about the details of competitive ones. In addition, research into the adoption of technological innovations in industry, by the Case Western Reserve University, USA, has indicated that 'follow up' or 'post audit' evaluations are usually found to be so biased as to render them of extremely dubious value anyway. Because of such a lack of published data, it becomes particularly vital to point out the general procedures which have, in practice, tended to prove most appropriate.

Direct factors

The traditional approach to capital investment tends to be to look at the readily apparent effects of the investment. Although in practice other factors are frequently very relevant to robotic installations, the 'traditional' direct factors are obviously of prime importance. These include the following (some of which overlap each other).

COSTS

System purchase price

The cost of the robot system itself is obviously a major factor. The actual value will depend not only on the cost of the robot or robots used but also on all the ancillary equipment. Such equipment will be very application dependent, and whereas over half the cost of an assembly system might be taken up with bowl feeders, conveyors and so on, a materials handling task might require no ancillary equipment at all.

The cost of the robot itself is, of course, very variable, depending on both its mechanical complexity and the sophistication of its controller. Recent SCARA-type assembly robots are available with four **degrees of freedom** and a very limited controller for only a few thousand pounds, but most robots cost several tens of thousands of pounds, with many of the heavier duty and/or high-sophistication devices costing over £100,000. Generally, the higher priced robots are considered cheaper to install, easier to program, and require less special tooling because they are more universal. As a very rough rule of thumb, the robot purchase price may be 30–60% of the cost of the installed robotic system, with the peripheral equipment costing 30–50%. Additional costs are engineering, installation and training.

Cost of special tooling

The cost of integrating system components such as conveyors and feeders may be substantial, requiring special tracking and the like to be custom made. Similarly, grippers are very rarely included in the cost of a robot, and, as explained in chapter 8, it is likely that a suitable design will nevertheless have to be customised for the parts it is actually to handle. Because of the uncertainty of costing all this additional expenditure, many suppliers offer **turn-key** systems, where, for the agreed price, they will provide a fully working integrated system. The term supposedly arose during the 1970s when, increasingly, rich customers of estate agents were in the habit of buying houses unviewed, which they expected to be decorated and fully furnished when they arrived, so that all they had to do was 'turn the key'. Engineering costs are typically 30–50% of the total cost of an installed system.

Installation cost

Although the cost of installation is sometimes charged directly to the robot project, in many cases factory layout has to be altered during a product change anyway, so that it would be unreasonable to charge more than the difference between the cost of this layout alteration and the cost of introducing the robots.

One time costs

In preliminary costing of a robot system, it is frequently forgotten that many of the workforce will need to be trained in various aspects of the robot's operation. Courses, which should be provided by the supplier, are required not merely for those supposed to operate and supervise the robots, but also for the programmers and maintenance personnel. Not only must this training usually be paid for (a limited amount is sometimes included in the robot purchase price), but those going on the courses (typically for a week) will be unavailable for other work. Experience suggests that 'informal' training – the supplier sending someone 'to show you how the robot works' – does not tend to be anything like as satisfactory as when they provide properly constructed courses. Installation and training can typically account for 10–15% of the total system cost.

Maintenance costs

Depending on the design of the robot and the conditions under which it is operated, varying degrees of preventive maintenance, regular overhauling and repair will be required. Modern designs incorporating sealed drive units tend to require less regular maintenance than older models, but for two-shift operation, maintenance may cost up to 10% per annum of the original purchase price.

Operating costs

Prime operating costs are usually caused by the cost of the power required to run the robot, controller and ancillary equipment. Despite a robot's inefficiency (compared to a human arm) the power costs are usually comparatively low; in the extreme, some assembly robots can be run off a single domestic-type power socket. In addition, of course, for certain tasks, consumables such as paint or adhesive may be far more efficiently applied, resulting in a saving when compared to other techniques.

Figure 16.2 Prime operating costs tend to be based on power requirements

Programming costs

Before the robot can perform the tasks required of it, someone must program it, and because it will perform those tasks repeatedly, it is worth making sure that the program used is the optimum – a small time saving on each cycle will soon build up. Textual programming can be a skilled job, and it is usually worth using somebody who has a flair for software, yet can work in collaboration with someone who fully understands the task being robotised. Teaching-by-showing, on the other hand, will tie up the best manual operator, and because of the unfamiliarity of the robot together with problems of editing, he may be occupied for substantially longer than might be expected.

Depreciation

Depreciation is, in effect, a recognition of the fact that the potential resale value of a machine becomes less as it gets older (just as a car does). Consequently, it is necessary to reflect this 'depreciation' in the company accounts if they are to continue to provide an accurate picture of the firm. This notional allowance for wear, tear and obsolescence does not involve any money transactions or **cash flows** (physical payments into or out of the company), it is purely a 'transaction' *within* the company. There are various methods by which accountants can choose to depreciate machinery, the most common being **straight-line depreciation** in which the depreciation allowed for each year is the same throughout the life of the device. Nevertheless, this is merely one of a series of equally valid yet arbitrary policies which a given company can choose, and although in some cases the depreciation may realistically represent the deterioration in resale value of a machine, with other methods the relationship is often quite tenuous.

A machine should be depreciated over its useful life, and for a robot in a 'low sophistication' application, this may be over ten years (some existing robots are still working after well over a decade's continuous use). Although robots in more technically complex applications may physically last over a decade, it is likely that technically they become obsolete before then. In many countries, tax schemes encourage depreciation as rapidly as is legally permissable, so that tax savings (which are allowed on the machinery depreciation) can be obtained as early as possible.

Cost of capital

It may be that in order to finance the purchase of a robot system a firm has to borrow money. There are various sources, but each will charge interest which will increase the cost to the firm of the purchase. In addition, as explained in later sections, the interest rate is of importance in evaluating projects, because the firm always has the alternative option of leaving its money in the bank and so gaining interest on it.

SAVINGS

Sale of old equipment

It may be possible to sell the old equipment the robotic system replaces (if any).

Government grants

Increasingly, governments are appreciating the advantages of investment in robotics and are offering incentives to encourage industry to automate using robots. In the UK, 50% of the cost of feasiblity studies into robot applications is recoverable from the government, and a grant can be obtained for up to a third of the cost of actually going ahead with the application – covering the cost of the robot and associated equipment, as well as development costs.

Labour savings

As mentioned above, although a firm may be cursorily interested in removing its workforce from an unpleasant or monotonous job, a prime motivator for introducing robots is usually the potential savings which can be obtained from not having to pay the wages and benefits of displaced human labour. In cases where a single robot can operate more than one shift, then these labour savings will proportionally increase.

The number of humans displaced by a single robot obviously varies, especially depending on what application is being considered, as it is rare for a robot application to be totally unmanned, without even limited supervision. Thus, although spot welding may require hardly any human intervention, resulting in an almost one-to-one replacement, arc welding may, on average only replace 75% or maybe 50% of a human worker per shift. Assembly is such a wide-ranging application that it is difficult to estimate how many humans a robot can replace – in some cases the robot may totally take over from the human, while in others it may replace less than 50% of a man.

Nevertheless, these are only the direct human displacements for a single shift. As mentioned in chapter 13, the introduction of comprehensive robotic automation may have major repercussions in the labour requirements of a factory – resulting in an apparent displacement of *several* humans 'per robot' introduced within the overall scheme. The Aston study, mentioned in chapter 14, considered that the average net effects were in the region of 2.5 people per robot, although other estimates vary from 0.8–6.2 jobs lost per robot installed. (This contrasts with the estimated 0.8 jobs *created* in robot manufacturing and servicing.)

Increased throughput

In some cases, a robot may be able to surpass the thoughput of a human operator. In these cases, the resultant improved utilisation of other equipment may also be of great benefit. In estimating throughput, the downtime of the installation is frequently overestimated. Whereas a breakdown in a transfer line may result in the whole line stopping, with a robot line the inherent flexibility of the robots can be used not merely to change from product to product, but also to take over from another temporarily inoperative robot further up the line. If this is not feasible, then the task can usually be continued for a short time manually, with a human stepping in while the robot is repaired or replaced by another robot loaded with the program of the original machine.

Consequently, whereas downtime of an automated line might be up to 25%, for applications in which replacement robots can be quickly reprogrammed to continue the task (such as welding or spraying) downtime using robotic systems may be only a few percent. Consequently, the nominal capacity of the installed robotic system can be less than that of the equivalent hard automation, because it will approach its nominal throughput more closely.

The integration effect

In addition to all the above 'direct' savings, there may be many other 'indirect' savings which are frequently neglected or omitted in justification analysis. Although these savings are more difficult to identify and quantify, they can be extremely significant. As computers integrate more and more of the factory environment, so the previously distinct processes within that environment likewise interlink. Whereas in the past it was legitimate (because it was a good approximation) to treat each process as a self-contained 'profit centre', which could be substituted largely in isolation for an alternative process, now, with increasing integration, it becomes progressively difficult to improve one part of the overall system without repercussions elsewhere.

Thus, a single robot operating as a direct replacement for a human, with no other alterations to the factory, may have negligible 'hidden' effect on the organisation as a whole (and consequently may be utterly uneconomic!). With several integrated robots, however, there are immediate implications for predictability, scheduling and so on, which may cause major benefits elsewhere in the factory if full advantage is taken of them. Communication brings integration with other processes, which allows increased efficiency and competitiveness overall. It is no longer then valid to conduct an economic evaluation merely of the direct savings of a robot system – as the level of integration with, and so the impact on, the remainder of the factory is increased, so too is the importance of the indirect-factors in any economic justification. This **integration effect** also forces the consideration of strategic as opposed to solely tactical factors in any economic evaluation of robotic systems. This, of course, suggests the need for a detailed understanding of the potential consequences of high technology by those most able to appreciate its likely strategic implications, namely top management (as mentioned in chapter 14).

An analogy to the integration effect is the ancient story of the Tower of Babel. In the story, the construction of the huge tower, employing a vast number of people, proceeds incredibly well until suddenly all the workers start talking in different languages. Progress effectively stops. The current movement towards factory integration is the inverse process to that in the story. As shown in figure 16.3, the twin advantages of robotisation are increased universality (providing the flexibility for reuse in different jobs) and increased intercommunication (resulting in integration with other sytems). Both these should result in greater *overall* efficiency and competitiveness, and hence increased financial benefit.

It is as if some new workers were taken on at Babel who could each

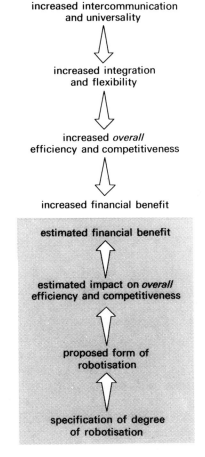

Figure 16.3 The integration effect

tackle several jobs and who could all speak the *same* language. In deciding how much they were worth, it would be wrong to consider merely how well they could do the specific job they were involved in at any one time. Their greatest contribution would be the 'hidden' benefits brought about through greater integration (that is, they could work *together* more efficiently than others) and their abilities to keep changing tasks (which allows them to cope with several jobs). The decision to employ such workers is more than simply a tactical decision ('Is it cost effective for them to do such-and-such a specific job?'), but is in addition strategic – employ enough and the tower may get built after all!

In the same way, evaluation of the benefits of robotisation *requires* investigation of the hidden benefits. It must be stressed that this is not merely because using traditional criteria projects may be rejected and therefore additional savings have to be found somewhere! It is instead because robotisation *changes* the level of integration, and therefore the 'spread' of benefits. Tracking down these hidden savings is simply a recognition of the increased sphere of influence of a robotic system over a conventional one, and therefore of the wider *coverage* of savings incurred by it due to the integration effect.

Indirect factors

In some cases, of course, it may be necessary to employ 'educated estimates' for indirect savings, which leaves one open to criticism and stubborn resistance from others in the firm. Nevertheless, if these savings are disallowed then they are effectively being treated as negligible (which they are not) and this is likely to be far less accurate than a realistic estimate. Some sources of indirect savings are as follows.

WIP inventory

If the use of robots results in a more efficient process, then parts will be produced faster and more consistently than before, possibly using a reduced number of discrete operations. As parts pass through the factory more smoothly than before (so incurring less overheads), fewer parts are actually 'tied up in the system' at any one time. As a result, the **work-in-progress** (WIP) inventory will cost less, so less of the company's capital will be tied up. This can be a very major factor indeed.

Raw materials inventory

More consistent manufacturing cycle times reduce the need to carry large 'buffer stocks' of raw materials. It may also be that employing a robot allows the use of lower cost materials, or maybe cheaper forms of fasteners, again reducing capital tied up in the raw materials inventory.

Finished inventory

More efficient processing results in lower unit costs, resulting in reduced inventory cost. In addition, the lead time to change to a new product using the flexible robot system is short, and the actual production time itself may also be shorter than before. This means that a firm can get away with maintaining a far smaller finished inventory as a buffer against sudden orders, because should a large order come in, the factory can rapidly respond and manufacture the required products. The savings incurred by reducing (or even eliminating) this 'safety stock' can be significant, especially as the consumer demand for wide product-style diversity is steadily increasing. Likewise, the more robotised the production process, the more cost effective it becomes to produce very small batches of a particular product.

Scrap and rework

Programmed operation, once the first run has demonstrated the validity of the program, eliminates the potential for operator error. Similarly, because inspection facilities are usually incorporated in some way into a robot system, the moment that a part is produced which is out of tolerance, a human supervisor can be summoned to correct the fault. In this way only one part may need be reworked or at worst scrapped, as opposed to a whole production lot. The costs of the inspection are easily paid for out of the reduction in wasted parts. In certain applications, such savings can be substantial. Naturally, reduced reworking also improves productivity.

Inspection

As explained above, it is frequently economical to automate inspection as an integral part of a robot system, and so the requirement for manual inspection of parts (at best a tedious and inconsistent operation) may be eliminated.

Operator training

Due to job turnover, training is a continuing cost. Because a robot operator is largely in the role of supervisor, the complexity of the training can be less than if he were actually to perform the task himself, and so the time required is reduced. In addition, one human can supervise several robots, and anyway job turnover (requiring training of new personnel) may be reduced if an originally unpleasant job is improved by robotisation.

Materials handling

The cost of running a forklift truck, together with paying a driver, is high, yet many studies have shown that parts may spend up to 95% of their processing time either being transported from operation to operation or else waiting in queues, and that an average part being processed in an average manufacturing plant spends only 5–15% of the time actually being worked on. Robot systems may combine or eliminate operations, so reducing materials-handling costs. As it is the introduction of the robot which is directly responsible for these savings, they should be included in its justification evaluation.

Floor space

In many cases, the introduction of robots requires less floor space than alternative methods, because of the consolidation of machines, less buffer stocks of components between stations, less need for inspection stations and so on. Provided that the floor space saved is subsequently utilised productively, then it can be classified as a first-year saving for the robot system.

Quality

Although difficult to quantify, quality improvements may result in savings other than those due to reduced rework and scrap. It may result in lower warranty claims (and so lead to a reduction in the cash reserve needed to cover such claims), less lost time in operations further down the line, customer 'goodwill' and increased repeat orders. As well as increased quality, introduction of robots may also result in greater quality *consistency*, allowing more reliance to be placed on quality predictions.

Product change

The inherent flexibility of the robot system may allow far faster adoption of new product designs, allowing a firm to make a more rapid entry into a new market than its competitors, so giving it a market lead. This is the kind of strategic consideration which may be a decisive factor in choosing to robotise.

Utility costs

Large-scale introduction of robotic systems may allow the firm to spend less on 'utility costs' relating to human comfort, such as heating, lighting and ventilation. In applications such as spray painting, savings on forced ventilation can be considerable.

Safety

Although, as noted in the last chapter, robots bring their own safety hazards, they can also, of course, remove humans from risky environments. In certain cases, it may be worth considering possible savings on personnel injury claims, lost time costs due to injury, disability or even death benefits.

Labour unrest

It may well be that elimination of unpleasant and/or unsafe work may improve the attitude of the workforce, resulting in fewer strikes and lessening any general unrest. Similarly, elimination of humans from a plant avoids problems of uprooting families should the plant be required to move location, and the cost of redeployment. Nevertheless, quantifying such benefits, let alone placing a monetary value on them, is notoriously difficult, and it is probably best merely to bear such factors in mind (if they seem particularly pertinent) when conducting an evaluation, rather than try to actually include them in the economic assessment. Even attempts to quantify the likely benefits from such factors as enhanced flexibility may be extremely difficult, as insufficient evidence may be available to estimate future net returns. Nevertheless, such estimates are, of course, vitally necessary for a comprehensive evaluation.

Payback

Of all the different investment appraisal methods, **payback** is the most popular, largely because it is so simple to understand and apply. Unfortunately there is no universally agreed definition, but in principle it is the time taken for all the various incomes resulting from a project to equal (or 'pay back') the original expediture. Consequently, in measuring the cash break-even point, payback is really more of a measure of **liquidity** (the ability of a firm to find cash at short notice to meet any business eventuality), rather than of economic efficiency. Where payback may be of value is in areas of rapid technological change (as with some but not all robot applications), where it can indicate the period during which the investment is at risk. If robots in a given application are to be replaced due to technical obsolescence long before their useful working lives, then it is useful to have an indication of whether the initial investment will have already been paid back.

This does *not*, however, mean that payback is a useful or even valid method for *selecting* projects. If we look at the three projects in figure 16.4 for instance (where by convention a negative cash flow represents money invested in the project and positive flows indicate earnings from the project), blind use of payback would suggest that the first two projects were of equal merit and superior to C, despite the fact that C recovers almost all of its outlay by the end of the first year. Similarly, what if project C has some substantial cash flows occuring *after* the payback period? Payback simply does not take account of the total useful life of a project.

When comparing potential robotic projects with manual equivalents, the inability of payback to incorporate the probable life of the project is likely to favour manual approaches, as returns from automation tend to be in later years than with manual equivalents. If steady increases in demand for products is expected, then this puts robots at an even greater disadvantage using payback, because manual methods would primarily involve increasing costs, whereas the robot system would be largely fixed from the outset. Similarly, lack of inclusion of both the project life and product plan is unlikely to reveal the substantial advantages of the inherent flexibility of robotic systems in accommodating the requirements for frequent introduction of new products into a volatile

project	initial cash flow	positive cash flows each year			payback period in years	rank
A	−27 500	25 000	2500	15 000	2	1 =
B	−27 500	2500	25 000	15 000	2	1 =
C	−27 500	22 500	—	20 000	2¼	3

Figure 16.4 Three projects ranked by payback

market. In addition, some types of project demand substantial injections of capital later in their lives, requiring multiple payback times, so confusing the whole process of using payback as a means of selection.

Thus, payback is seen not only to ignore completely the value of cash flows of projects which 'mature' after the payback time, but also to be incapable of selecting between the economic merits of competing projects. In addition, it suffers from the shortcoming that is does not cater for the fact that a firm always has the alternative option of placing cash on deposit in the bank. The longer the money is in the bank the more interest it earns, so from the firm's point of view a sum of money now is potentially worth more than the same sum in a years time (because in that year the first sum has earned the firm some extra cash in the form of interest). This 'time value' of money is dealt with further in the section on net present value.

Some supporters of payback suggest that because forecasting is at best imprecise, then it is valid to only look forward a small number of years (usually between one and three) by using payback. In other words, they are attempting to use payback to deal with the problem of risk. The higher the risk the shorter the acceptable payback. In reality payback is far too crude a tool for the problems of risk (a more suitable approach is covered in a later section), especially as it in no way reflects the existence of risk up to the payback time! Others accept that payback is really a measure of liquidity, but misguidedly argue that it is this that in fact makes it a valid tool, because it is liquidity problems that often prevent a company investing in seemingly profitable projects. Although liquidity considerations are obviously of importance to a firm, there are far more appropriate methods of solving such problems than by sacrificing potentially lucrative projects through inappropriate use of payback.

Return on investment

Return on investment (ROI) is one of several different 'rate of return' measures, all of which suffer from the same major drawbacks. ROI is generally considered to be the percentage ratio of annual profit from a project divided by capital cost, but there is no single definition for the profit and cost. Different firms use various combinations of tax, interest charges and depreciation to arrive at a figure for the profit, and of course, depreciation itself can be accomplished in a variety of equally valid forms. Similarly, some feel that only the first year's profit should be used, while others employ the average annual profit. Clearly, the whole procedure is somewhat arbitrary.

Not only do the measures not reflect the useful life of a project or take account of the *timing* of cash flows (so failing to reflect the advantages of near flows compared to distant flows), but the 'magic' percentage figure obtained can be radically altered merely by a change of depreciation policy (which is arbitrary anyway). Indeed, the ROI approach to financial decision making and control was originally pioneered by Du Pont as a way of approaching the problem of top-level control, rather than as a magic formula for solving these problems.

Net present value

Neither of the above two methods take into account the 'time value' of money which arises because a given sum is worth more to a firm now than if it arrives at any time in the future, as in the interim period the money can be put to profitable use (if only by earning interest in the bank). For example, if the bank interest rate is 10%, then £110 received in one year's time is the equivalent of £100 received now (and then put on deposit). More generally, assume that the money can earn interest at a rate of r per cent per annum. If the firm places P pounds on deposit, then it can be shown that in n years time the deposit will have grown to:

$$P(1 + r)^n$$

Alternatively, this can be rewritten to show that the 'present value' P of a sum of money S arising in n years is given by:

$$P = S \times \frac{1}{(1 + r)^n}$$

In other words, the value of any cash flows that will arise in the future can be **discounted** (using the appropriate **discounting factor** $1/(1+r)^n$ for the year in question) to obtain their present equivalent values. By using such **discounted cash flows** (DCF) for all of the cash flows in a project, it is possible to view the project in terms of a series of equivalent present values; that is, at last 'like is being compared with like'. In this way it is possible to compare legitimately cash flows due to occur in five years with cash flows occurring immediately, because all have been translated into their equivalent 'present-day values'.

If all the various discounted cash flows are added together (with negative values corresponding to outgoings and positive to earnings) then the resultant value will be the net cumulative discounted cash flow, or **net present value** (NPV), for the whole project. This value will indicate whether the project is worth considering or not, because if the NPV is zero then in present value terms (as opposed to the actual nondiscounted values) the project just breaks even, the firm neither benefiting nor suffering from investment in the project rather than leaving the money to gain interest in the bank. If the NPV is greater th?n zero, then the firm will benefit from a net gain if it invests in the project; if the NPV is negative, then the firm will earn more if it leaves the money in the bank.

The NPV of different mutually-exclusive potential projects can be used as figures of merit for each one: the project with the most positive NPV is predicted to be the most profitable. If we employ the NPV technique to rework the three-project appraisal used before, then it can be seen in figure 16.5 that project C is now appreciated to be far more profitable than project B, and only marginally worse than A. At last the relative timings of cash flows are being taken into account. In addition, of course, discounting implies that cash flows a long time in the future have a relatively small impact – thus predictions of distant events (which are bound to be 'suspect' anyway) are seen automatically to be taken into account less than those of more accurately predictable events.

There are those who are suspicious of NPV for use with robots, because it seems to favour projects which yield a relatively quick return, which robot systems, of course, may not. These people, however, miss the point. It *is*, ideally, better to obtain quick returns, because they are worth more! That, however, does not mean that, using NPV, upgrades of existing equipment which offer such quick returns will always appear preferable to introductions of new robot technology offering later profits. On the contrary, provided that an *overall long-term* view is taken, the competitive advantages of robotisation may far outweigh the alternative short-term performance of a given profit centre which is merely 'upgraded' using conventional equipment. What is required is

project	initial cash flow	positive cash flows each year			net present value (r = 10%)	rank
A	−27 500	25 000	2500	15 000	12 170	1
B	−27 500	2500	25 000	15 000	10 125	3
C	−27 500	22 500		20 000	11 530	2

Figure 16.5 Three projects ranked by net present value

that the indirect factors mentioned earlier (such as flexibility to product changes) are fully included, so that the wider benefits of a project are appreciated. Thus, whereas it *is* true that discounting procedures used on their own may not be suitable for robot economic evaluation, they *are* ideal when a comprehensive approach is adopted by including both direct *and* indirect factors over the long term.

Internal rate of return

An alternative form of discounting commonly employed involves a trial-and-error approach. Successively higher discount rates are applied to a project (that is, the supposed rate of interest which could be obtained with the money in the bank is increased to maybe 15%, then 20% and so on) until a rate is found at which the project appears to break even, with a NPV of zero. This then implies that the final value obtained for the interest rate (say 25%) is in fact the **internal rate of return** (IRR) for the money invested in the project. In other words, investing in the project would be the equivalent of investing the money in a bank which was offering (in this case) an interest rate of 25%! The value of the IRR (sometimes alternatively known as **yield** or **discounted cash flow rate of return on investment** – DCFRR) allows projects to be compared directly against the interest rates obtainable from alternative forms of investment.

As with NPV, IRR calculations do not rely on arbitrary accounting conventions, but concentrate on profitability analysis of the predicted incremental cash flows. For a project to be of interest, its IRR must be higher than the company's cost of capital, yet in a period of economic uncertainty the bank lending rate is likely to fluctuate. By seeing just how much leeway the project has if interests in fact change, it is possible to incorporate a degree of risk protection into one's appraisal. Such **sensitivity analysis** is just one method of dealing with the problem of uncertainty and risk; in its simplest form the technique involves varying, to reasonable degrees, different parameters *one at a time*, to discover the sensitivity of the project's NPV or IRR to the variations. In this way questions such as 'What if the predicted cost of the project is inaccurate by 10%?' can be investigated. Modern microcomputer programs are available which allow such investigations to be conducted very easily. Using IRR to rank projects can in practice be problematic under certain conditions, but the cumbersome approach that must sometimes be adopted is frequently considered worthwhile because the IRR percentage figure is conceptually so easily understood.

Guidelines

In conducting an economic analysis of a potential robot application, above all the robotisation must be considered as *strategic* (involving longer term competitive advantage) rather than solely *tactical* (short term profitability of a specific profit centre). So, as stressed in chapter 14, support for the project must come from senior management who can truly appreciate the global company benefits of a proposal. In addition to this overriding consideration, however, there are various guidelines which can also be adopted. Firstly, leasing should always be considered as an alternative to actual purchase of capital equipment. Not only are a few robot producers actually leasing out robots already, but most of the banks and leading financial institutions now have a leasing department. With leasing, the lessee (the user) pays a regular sum which is adjusted to take account of the capital allowances which the lessor (the owner) is able to claim on the leased equipment. In addition, no deposit or down payment is required, and payments can often be varied to suit a business' cash-flow position. At the end of the initial leasing period the lessee is normally entitled to continue to lease the equipment at a nominal rental only.

Naturally, the cost justification process becomes far simpler when leasing is concerned, as the lessor in effect deals with depreciation, cost of capital, and so on. Typically, however, the lease is only for 5–7 years, and the robot may be depreciated over a period even shorter than this. Nevertheless, if the robots are to replace manual workers, then the form of paying for the two competing approaches is very similar, again making evaluation simpler. Finally, leasing may mean that it is possible to avoid the need to apply to the capital appropriations committee, so streamlining the whole acquisition process by avoiding the otherwise necessary bureaucracy. Nevertheless, this may be a mixed blessing. The Aston study mentioned in chapter 14 indicated that unsuccessful attempts at robotisation had on the whole, few problems obtaining finance – suggesting that restrictions on capital forced closer scrutiny of projects, in turn leading to better final proposals.

When a company is considering the first introduction of robots, it may well be quite legitimate not to worry too much about their cost effectiveness. For a start, it is always going to be cheaper to run several robots rather than just one, because people such as the maintenance operator are required whatever the numbers, and his cost can be spread out over all of the robots when there are more than one. Likewise, there is always a learning curve with such radically new technology, and one way or another a company has to pass up it.

It is frequently wise for a business to consider these first few robot applications as something of an experiment – not so much to prove whether robots can improve productivity (that has already been demonstrated) but more to build up some in-house experience in the field. Better to pass along the learning curve when the economic survival of the company does not depend on it. That said, it is important to choose these early applications carefully. There are any number of suitable tasks for which robots have been proved ideal and it is best to choose from among these (figure 17.6). The more adventurous and ambitious projects can come later – human nature being what it is, if the early projects do not work well, there may be no further opportunity for any others!

Ideally, robots should be utilised 24 hours a day, but if this is impossible then two shift operation should be used. However, this does not imply that output should necessarily be much higher than before (assuming only one shift was worked previously). It may be feasible to arrange that fewer robots are bought, so that their eventual output is only slightly

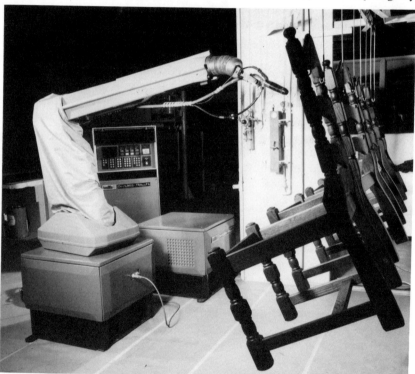

Figure 16.6 Surface coating – a well proven robot application

higher than obtained previously. Where it really is impractical to operate robots for more than one shift, then it should be arranged that they
work unmanned during lunchbreaks and for as much 'overtime' as
possible. Under such circumstances, however, it will be far more difficult to make the systems cost effective.

Naturally, many readers actually considering installing robots may
feel somewhat overwhelmed by all the problems of properly justifying
their introduction. What at first seemed like a relatively straightforward
proposition may by now have appeared to have assumed awesome
complexities! It is worth the trouble. If your company accepts the robot
proposal using nothing but simple payback techniques, then well and
good (although with such an economic selection strategy running your
firm it might be worth looking for another job before the rapidly
changing market forces take their toll).

If, on the other hand, a simple assessment of the project does not
satisfy the company requirements, then explain the implications of the
integration effect and investigate the potential 'hidden' benefits brought
by the introduction of the proposed system. Also attempt to indicate (if
necessary) the advantages of employing discounting procedures in order
to obtain a true picture of a project's cash flows. The uphill struggle will
certainly be worthwhile if, as a result, the firm introduces strategically
profitable robotic systems which it would otherwise have considered
uneconomic. The robots involved will have been painstakingly designed, optimised and developed by the robot supplier; it is only
appropriate that a proportional effort should have been made by the
purchaser to ensure that the robot was installed in the most cost-
effective application.

Part V
CURRENT PROSPECTS

Throughout the world, robotics is moving forwards extremely rapidly.
The two chapters of this final section of the book cover some of the
research work which is fueling that progress, as well as both forecasting
the direction such progress will take and indicating some of the predictions which various others have made. Of all the new technologies,
robotics may eventually have the greatest impact – it is as well to have
an idea of where it may be going!

17

Something's cooking . . .

PRESENT RESEARCH

Robotics, together with such things as biotechnology and information technology, is one of the new 'sunrise industries' which are seen as potentially having major influences upon the development of mankind, and that are now only just 'dawning'. Understandably, research in such areas is frenetic, and robotics is no exception. This chapter deals specifically with such work, yet it would be impossible to do justice to all the branches of robotics research currently being pursued; it would be difficult even to cover them all.

Instead, this chapter attempts to provide a 'taste' of the exciting research being conducted throughout the world. Much has necessarily been omitted (although some of these areas have been covered elsewhere in the book), but the topics that are covered offer something of a cross-section of the activities of both academic and industrial researchers. Some topics are, in reality, very major areas of research; others are, in practice, quite isolated aspects of research. Yet not to include such areas would rob the reader of a feeling for the richness and diversity of work currently being pursued, not just towards industrial robots but towards nonindustrial robots as well.

Much fundamental research is directly robotics related, but some, equally influential, is of more general application, as is the case with computing science. Of the work being conducted into robotics applications, some is to develop completely new applications while the rest is to improve existing ones. Yet for those conducting the work, whatever the actual field of robotics research, there is a deep-seated excitement in knowing that one is 'pushing back frontiers' by working at the very boundary of man's knowledge. Despite the necessary humdrum activities of everyday work, there is an undeniable thrill in sometimes stopping for a moment, and suddenly remembering again that one is helping to turn science fantasy into reality . . .

Arm design

Alternatives to the more 'obvious' arm designs are a potential source for significant improvements in performance and flexibility. The recent SCARA-type assembly arms are an example of such a rethink. University College, London, is investigating the possibilities of a continuously jointed arm with a very large number of independently controlled degrees of freedom, while Chalmers University, Sweden, has conducted several years work on a paint spraying robot based on the human spine.

This 'Spine Robot' (figure 17.1) looks rather like an elephant's trunk, and has a working envelope which is almost a perfect hemisphere 4 m

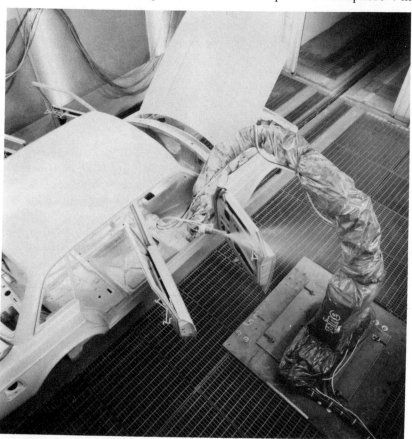

Figure 17.1 The 'Spine' robot, developed at Chalmers University, Sweden

high and 3 m in diameter. The arm is actually made of two spines (arranged serially) consisting of 100 ovoid discs whose faces roll on each other, connected by four steel cords. Movement is achieved by pulling the cords using hydraulic actuators, and this allows seven **degrees of freedom** permitting, for example, the tool to be held stationary while the arm itself moves.

Other alternatives to arm design include using special light-weight materials for construction, as well as using optical tracking techniques for accurate location of the end-effector independent of any axis encoders. Such approaches may allow arms to be constructed with far higher apparent stiffnesses than before. At Stanford, USA, research is being conducted using an arm designed deliberately to be extremely flexible in one lateral axis, yet rigid in the others. Encouraging attempts have been made at driving it from one end – rather like a fly fisherman accurately placing his fly with only a flick of the wrist. High arm speeds will also be required in the future and MIT, USA, are aiming at accelerations of three times that of something falling under gravity, and speeds of 5 m/s with payloads of maybe 10 kg

Artificial intelligence

After a period of under-funding, it is now being realised that **artificial intelligence** (AI) is, in fact, likely to make one of the most important contributions to future robotics, and indeed it is fundamental to the fifth-generation computer project' (covered in a later section of this chapter). A lot of misunderstanding has arisen concerning AI – 'Surely a machine cannot be intelligent?' Perhaps one of the most convenient definitions of AI, however, is that it is the science of making machines do things which, on the whole, if performed by humans, would be attributed to intelligence. It is as pointless to become paranoid about the use of the word 'intelligence' when applied to computers as it is about using the word 'flying' for both birds and aeroplanes.

Because some AI systems can already be made to adapt themselves, it is possible, for example, to make them learn from their own mistakes. As a result, the old adage, 'Computers can only ever do what they are programmed to do', is not strictly applicable to this kind of system. A computer can, for instance, end up improving its own program to a standard that no human could have accomplished. It is *not* science fiction to assume that within the foreseeable future a computer system may be constructed which appears to match the human brain – at least in particular areas.

Such a machine will pass the so-called **turing test**, proposed in 1950 by the British mathematician Alan Turing, a pioneer in computing, as a criterion for judging 'mechanical intelligence'. The test is a hypothetical procedure based upon interrogation, in which the human interrogator can communicate with another human and a computer via a terminal. He does not know which is which, and has to decide on the basis of the answers given. If it is impossible to tell the difference, then, to all intents and purposes, the computer may be said to be acting 'intelligently'. In effect, the test encapsulates the concept that if, in practice, an imitation is *completely* convincing, then there is no point in attempting to differentiate it from the original by 'splitting hairs'.

It has been pointed out that even the claim, 'No computer will ever be able to take decisions based on intuition,' is probably little more than an attempt to calm public disquiet about computers. Although one can, of course, define intuition as a purely human attribute, this begs the question, 'Is it possible for a computer to take decisions which, *if taken by a human* would indicate the use of intuition?' – to which the answer is 'Why not?' Very few authorities in AI believe that there is anything fundamental about computers to stop them eventually behaving in this way – although this is, of course, a highly contentious issue. Meanwhile, research continues throughout the world. The potential impact for robotics of such work is covered in the next chapter.

Yet, already there are many practical developments from AI research, such as the emergence of **expert systems**. In such systems the knowledge of human experts in a particular field is reduced to a set of rules, including such less precise forms as 'rules of thumb'. A computer can then employ this **knowledge base**, using an **inference engine** to work out the logical consequences of a set of rules taken together. One of the earliest expert systems was 'Mycin', developed to help in diagnosing infectious diseases such as meningitis. Since then there have been many successful systems, including 'Prospector', designed by the Stanford Research Institute to locate mineral deposits, which actually managed to locate a $100 million molybdenum deposit at Mount Tolmen, Washington State, USA, which the human experts could not find. Expert systems are being heavily researched in the fifth-generation project and are likely to become extremely important in robotics work as a means for robots to learn certain more advanced skills.

Assembly

As robotic assembly has been singled out as one of the most important applications for second-generation robots, it is not suprising that research work in this field is being conducted at almost every major robotics centre, yet there are so many aspects to robotic assembly that the range of this research work tends to be very wide. Investigation into the special design of components and systems suitable for robotic assembly is being conducted at such Universities as Salford, UK, and Massachusetts, USA, while suitable task-analysis techniques are being developed at Purdue, USA.

Work on part-mating and compliance is particularly strong at the Charles Stark Draper Laboratory, USA, while investigation of multiarm systems is conducted at several research institutes, among others at IPA Stuttgart, Florence, Pisa and Milan. Parts feeding is being heavily investigated at the Swedish Institute for Production Engineering Research and at the University of Salford, UK, who have designed a form of linear feeder suitable for the comparatively slow pace of robotic assembly. Meanwhile a methodology for cost-effective assembly is being devised at Imperial College, London.

Many companies are obviously researching strongly into robotic assembly, including several Japanese firms, such as Hitachi, and also IBM, who seem particularly interested in the automatic assembly of light components. A more unusual application is that being investigated by Unimation in collaboration with Lansing-Bagnall, UK, and Hughes Aircraft – robotic wire harnessing. This entails threading various wires between different points of the 'wiring harness' used in many electrical appliances. Since the late 1970s, Westinghouse (which now owns Unimation) has been developing a large scale experimental FAS known as APAS (adaptable programmable assembly system) for electric-motor assembly.

Automated guided vehicles

Because of the inflexibility of traditional wire-guidance systems, advanced research is being conducted into free-roving mobile robots using internal navigation systems to maintain position fixes and employing elementary maps of their work areas. The development of such mobile devices involves research into sensors, navigation, control and communication techniques, many requiring novel software structures. Much work into mobile wheeled and tracked robots is being conducted in France, including that at the French National Institute of Applied Science, and on the autonomous robot 'Hilaré' (figure 17.2a) made at the National Centre for Scientific Research. In addition, of course, there is work being conducted on (*currently* less practical) legged systems, as covered in chapter 12.

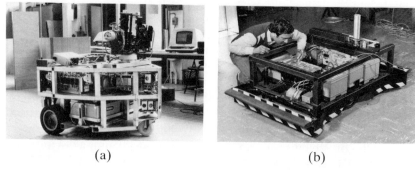

(a)   (b)

Figure 17.2 Two examples of mobile-robot research: (a) 'Hilaré',
National Centre for Scientific Research, France (b) prototype robotic
AGV, Imperial College, London

. Just as the Stanford Research Institute became experienced in mobile
research robots (such as 'Shakey') in the 1970s, so, under Dr Mikki
Larcombe, Warwick University, UK, has built up a reputation in free-
roving automatic trucks intended for industrial use in factories and stock
yards (mentioned in chapter 12). Such systems incorporate sophisticated
sensing capabilities to allow them to navigate through environments
such as warehouses. Trials are already taking place in a large empty
shop floor at British Leyland's old Triumph factory at Coventry, near
Warwick University – so truly robotic AGV may soon become commercially available. Research work on such systems is also active at Imperial
College, London (figure 17.2b).

Future systems may follow a buried wire for part of their journey,
then depart from it to execute a particular task, returning to the wire for
guidance or communication purposes. Sonar devices will provide information about walls and objects, and a further possibility is to employ
an upward-pointing photosensor to use factory lighting to confirm position. A permanent communication link with a central scheduling computer is likely to involve an infrared network (using the same priciple as
remote controls on domestic televisions). The Warwick AGV maintain a
computerised 'map' of an area which is checked against actual data from
a range of sensors. In this way a truck need not rely entirely on any one
source of information, and will not be disabled by a few missing or
confusing cues. It is intended that limited voice synthesis will be used by
the trucks to provide running commentaries – explaining, for instance,
that they are not out of control but merely executing multipoint turns!

Automated factories

In the early 1970s the Japanese government instigated the **methodology for unmanned manufacture** (MUM) project, with the intention of funding the development of a largely unmanned factory. Although the project has since been somewhat scaled down, it is still intended that a small factory will come into operation in 1985 for the manufacture of 15 different component types such as wheels and shafts for gear boxes. Modular construction will be employed as far as possible, and as well as conventional CNC machine tools, more revolutionary equipment such as lasers will also be used. Robots will be employed for loading and unloading of machines and conveyors. The system will also be able to deal with raw materials, **swarf**, and bought-out parts.

Fujitsu Fanuc, the world's largest producer of NC systems, has already built a largely automated factory costing £18 million, for the manufacture and assembly of spindle motors and both DC and AC servo motors. The factory (figure 17.3) consists of two stories, with the ground floor as the machine shop with 60 NC machines (32 loaded by robot), and the first floor as the assembly shop, with 49 robots arranged in 25 **cells** – though 35% of assembly is still currently performed manually. The two floors of the factory are linked by an automatic warehouse. Fanuc claim that introduction of the factory has cut manufacturing costs by 30%.

Much long-term research is being conducted in the field of automated factories. An example of this is 'CIDMAC' (the Computer Integrated Design, Manufacturing and Automation Centre) project at Purdue University, USA, which is one of the major thrusts in manufacturing automa-

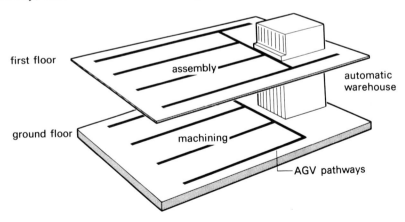

Figure 17.3 Schematic of the Fujitsu Fanuc automated factory

tion research. The goal of the project is to achieve true integration of the various technical areas that contribute to manufacturing productivity.

It is anticipated that over $100 billion will be spent between 1984 and the end of the decade on manufacturing automation generally. Some believe that this all appears to be leading towards the day in which **unmanned factories** consist of robots tending machines in windowless, unlit, unheated and unventilated buildings. Indeed, if the factories are truly unmanned, the 'ceilings' of the buildings may only be a metre or so from the floor! Some of the major components of automated factories, such as CADCAM, FMS, automatic warehousing and so on, are dealt with in subsequent sections of this chapter.

CADCAM

With the dramatic decrease in the cost of computing power in the 1970s, together with high-resolution graphic displays and advanced software, it became feasible to use computers in the drawing office for the preparation of technical drawings. Such CAD systems can increase the productivity of a draughtsman by three times or more. Although the systems are still largely used only by the bigger companies, their sophistication is steadily improving, and increasingly the systems are able to transfer engineering data automatically to external management control and information systems. With the high reliability and ruggedness of modern microelectronics it becomes possible to locate substantial computing power actually on the shop floor, so allowing the CAD systems to communicate directly with the CAM equipment in the factory. Such equipment might include CNC or DNC machine tools (ranging from lathes to milling machines), robots for loading and unloading them, as well as robots for performing actual manufacturing tasks.

Much research is being conducted by industry to develop the complicated software which will allow CAD and CAM systems to communicate with each other, to form true CADCAM packages. In addition, because of the increased usage of sensory feedback in robotic systems, it is becoming possible to inspect and test products automatically after they have been manufactured. When such an approach is incorporated as part of a CADCAM system, the overall approach is sometimes known as CADMAT, standing for computer-aided design, manufacture and testing.

Farm work

As robots become more sophisticated, so it becomes feasible to use them outside the structured environment of the factory. Much farm work is very labour intensive, and serious consideration has been given to robotising the milking process. Outside of the milking parlour, as free-roving vehicles advance in sophistication it becomes possible to produce driverless tractors. These can be used for sowing, ploughing, fertilising, crop spraying and so on, releasing the farmer for the increasingly technical job of caring for crops and livestock.

A programme at the National Institute of Agricultural Engineering, UK, investigated designs of driverless tractors (for ploughing) which did not make use of any buried cables, transmitting beacons or similar devices. An optoelectronic method was successfully employed for accurately following the line of the last furrow, although the problem of turning the tractor at the end of the field (the 'headland') was found to be more difficult. It involved marking the boundary of the field with reflecting posts and also employing ultrasonics to search for the last furrow. Such techniques are useful not only for ploughing but also for the other tasks mentioned above, especially crop spraying, where it is important that each pass should closely fit the one before so as to prevent gaps or double spraying – both of which can be harmful.

In addition to agriculture, both forestry and fish-farming are seen as potential users of robotics technology, and Japan in particular is researching into suitable approaches for such introductions.

Fifth generation computers

At the beginning of the 1980s, the Japanese launched the 'fifth-generation computer project', with the goal of developing, before the end of the decade, revolutionary new computer systems which incorporated and exploited the concepts of artificial intelligence. In a statement released in 1980, the Japanese Information Processing Development Centre, in explaining the deficiencies of current computers, said, '. . . what is required is, in a word, a computer system with common sense. Ideally, such a system would be able to draw related items from its knowledge base in making judgements concerning matters which it had not been given instructions about'.

In addition to AI software, such systems also require special forms of computer architectures. Europe and the USA are joining in the race to develop such fifth-generation computers, and the results of the work are likely to have an incredible impact on robotics. The first three years of the project are taken up developing 'tools' with which to realise the actual project goals during the remaining seven.

The project's technical goals include increasing the 'intelligence' of computers so that they can interface with humans using everyday natural language, employ knowledge-based expert systems to put stored knowledge to practical use, and actually exhibit the functions of learning, associating and inferring. Software generation will be far simpler, and the actual computers will be light, compact, high speed, large capacity machines. Realising these goals involves using new approaches to computer hardware as well as software: the conventional **Von Neumann-type architecture** is reaching its performance limit because such computers perform all their data processing in series.

Such a serial approach places a theoretical ceiling on the maximum speed of computation, governed by the speed the electrical signals can travel. Already modern 'supercomputers' have to be physically built so that no wire is longer than a few tens of centimetres – otherwise signals take too long to travel down them! The only practical approach for overcoming this theoretical limitation is to perform several processes at the same time, by employing a form of **parallel** architecture instead of the conventional serial one.

Of course, there is more to such work than simply running several processors at the same time and hoping to somehow split a problem up between them. One approach to this problem is the concept of **dataflow architecture**, in which operations on data are carried out immediately it is available, instead of waiting until a single control sequence gets round to each operation. MIT, USA, are currently building a 'connection machine' comprising *1 million* crude processors, each with about 100 bits of computer memory! The processors are being built on custom-designed chips, 64 per chip, so there will be 16,000 chips in the finished machine.

In the UK, at Imperial College, a fully parallel machine known as 'ALICE' (Applicative Language Idealised Computing Engine) is being constructed. The machine is made from only three kinds of custom-made basic elements, replicated over and over again. Most current microcomputers can process only about 250,000 instructions per second, yet even a *desk-top* prototype of ALICE will run at the equivalent of 5 million instructions per second or more   In the same way as current architectures are unsuitable for fifth-generation computers, conventional computer languages are inappropriate for the new machines, and so methods such as **predicate logic** programming are being evaluated. This, in contrast to conventional programming in which a computer is given a series of steps to follow, specifies statements and questions in a form very like classical logic. The most successful implementation of logic programming is currently the language **Prolog**, championed by Robert Kowalski of Imperial College, London.

It is generally believed that the goals of the fifth-generation computer project are indeed largely achievable, and of course, the moment improved computers are available they will soon become components of sophisticated robotic systems.

Flexible manufacturing systems

Early precursors of FMS first started to appear in the alte 1960s when a UK company, Molins, produced a new kind of machine tool called 'System 24' which, in attempting to bring small-batch production nearer the efficiency of mass production, was a forerunner of later FMS. Unfortunately, the development costs of System 24 were so high that Molins eventually had to write off the whole project in 1973 at a loss (then) of £5 million. It was left to other countries such as Japan and the USA to build up experience in FMS design, and now increasingly the FMS **cell**, consisting of a robot tending one or more machine tools and ancillary equipment, is becoming one of the 'building blocks' of the modern factory.

A normal FMS consists of several such cells linked together both physically and by computer. The ability of cells in such a system to 'share' information and be globally supervised by remote computers, allows an overall strategy for the FMS to be followed, resulting in a truly integrated system. Such an approach is increasingly becoming known as **computer integrated manufacture** (CIM). Of course, the more flexible an FMS has to be, the greater the number of different tools that will be required for machining. Thus, very high flexibility incurs a reliability problem in physical storage and transport of tools.

The Citroen FMS, at Meudon near Paris, has storage capacity for 700 tools – 50 in each machining cell and 600 in a special auxilliary store. This four metre high store contains a manipulator which automatically selects required tools and places up to 24 of them on to a portable carousel sitting on an AGV, which in turn delivers them to the specified machining centre. When it is remembered that not only must spare tools be stored (in case of breakage) but that whole backup systems are needed (because breakdown of the tool store could bring the whole FMS to a halt), it becomes clear that the tool handling system for a highly flexible FMS is likely to be a major component of the overall system.

Figure 17.4 A factory robotic assembly cell – leading the way for practical FMS?

The number of FMS systems in the world is already into three figures, and seems to be doubling every three years. One of the major driving forces for the introduction of FMS is their ability to reduce **work in progress** (WIP) while increasing machine utilisation. It has been estimated that about 75% of all metal-working production consists of batches of under 50, yet traditionally with such small-batch work a part may spend only 5% of the time on a machine tool, and even there is only cut 30% of the time. This, together with machines frequently standing unused, results in large amounts of capital simply lying idle.

Indeed, the British United Shoe Machinery Company, in Leicester, estimated that production of a typical part by stand-alone CNC equipment cost 2.5% more than by FMS, and produced 66% more WIP and 34% more inventory (based on a 17 hour working day for the CNC machines and 24 hours for the FMS). Similarly, a £6 million FMS system at Anderson Strathclyde, Glasgow, is expected to save £1 million in inventory, and time spent in the machine shop should drop from four months to four weeks   Almost every robot centre and producer is in some way involved in FMS research. One of the most ambitious schemes is the US Air Force ICAM (Integrated Computer-Aided Manufacturing) programme, which has the long-term goal of building a highly automated aerospace factory. In the UK, the 600 Group, at Colchester, have already developed, with the aid of a government grant, an integrated FMS machining system for small-batch production of general turned parts. The system is called 'SCAMP' (Six hundred group Computer-Aided Manufacturing Process).

It consists of nine machine tools with eight robots servicing them, and it makes use of an advanced vision system for parts orientation. Small batches of turned components are being machined, untouched by human hand, within three days, whereas in the past the same task might have taken two months and 50 handlings! The average cost of FMS projects obtaining grants from the UK government has been about £3 million each, and companies of all sizes have started installing them – including several firms employing under 200 people.

FMS **software**

There are many tasks for the computer on an FMS system, but among the most important is that of enabling a manager to specify orders and to schedule the system according to various priorities, so as to achieve the highest machine utilisation – given the dates the parts are due, and given the lack of availability of any machines which are 'down' for whatever reason. In addition, operators, if used, must be instructed to load new batches of raw material and remove completed work, and pallets must be directed to and from the warehouse and appropriate locations using either conveyor systems or automated guided vehicles.

Machining programs for CNC devices as well as programs for the robots must be stored centrally and then distributed by **down-line loading** to the appropriate **direct numerical control** (DNC) device when the arrival of a given part dictates. Software must additionally control warehousing and maintain stock records, as well as keep track of every component. Finally, the whole system must be constantly monitored so that any malfunctions can be rapidly detected and corrected.

Many approaches have been suggested for structuring the complex software needed, and a lot of research in this field is still going on throughout the world, particularly in the USA, UK and Japan. Most approaches involve some form of hierarchy of control, yet because of the complexity of the software, it is necessary to attempt to partition it further by modularising the whole package in some way. This means that development of different modules can proceed in parallel, changes and upgrades need only require local alterations, the system is more resilient to failure, certain modules can be largely duplicated if, for example, an extra machine tool is added to the system, and the whole package is much easier to write and test. Nevertheless, the actual choice of the boundaries of the modules depends on which of the many robotic software-design strategies (adopted from the field of computing science) is chosen.

Despite such approaches, FMS software remains extremely complex to write. A recent FMS ideally required over £1 million of tailor-made software (over 40 man-years of work). However, because this would have taken about three years to write, while the mechanical equipment could be installed and commissioned in only 15 months, it was eventually decided to employ a much simpler software package than would have been ideal. As a rule of thumb, the computing side of an FMS may constitute 30% of the total cost, while the software alone may be 20% of the whole. Further, Ingersoll Engineers have estimated that only 10% of FMS built in Europe and the USA have been a complete success, with the majority of the remainder being problematic because of software flaws.

Fun

A great deal of robotics research work is intended to be of immediate practical benefit, improving on robot design or solving an application problem. A few, on the other hand (such as the Waseda University five-fingered robot, figure 17.5, which plays the University hymn on an organ!), while providing the researchers involved with experience in robotics, and probably indirectly solving a number of potentially useful problems as well, have as their prime goal the construction of a robotic device for which there is absolutely no requirement at all! The results, however, tend to be fun.

Probably one of the most gloriously unnecessary of such devices is the self-contained robot developed by Battelle, Pacific Northwest Laboratories, to unscramble a 'Rubik's Cube'! The cube can be placed in the

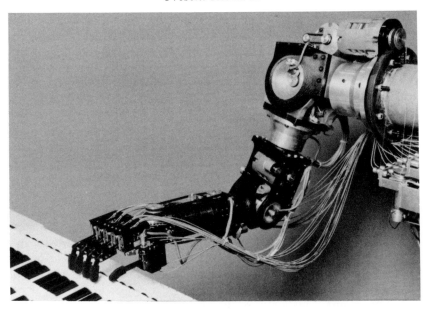

Figure 17.5 The organ-playing robot arm built at Waseda University, Japan

hands of the robot (designated the 'Cubot'), and the system then proceeds to read the colour patterns of the scrambled cube faces. Having determined the condition of the cube, the robot uses a special algorithm to determine the sequence of moves necessary to return each face of the cube to a single colour, and then, using its manipulators, it executes each move of this stored solution until the puzzle is solved. Although the length of time taken to solve the cube depends on the complexity with which it has been scrambled, even the most involved case takes less than three minutes! This must surely be the executive toy for 'the man who has everything' which beats all comers . . .

Future-generation computer components

As mentioned in chapter 2, ever since the mid-1960s the memory density of computers has approximately doubled every year, with rapidly falling power requirements (figure 17.6). If this rate continues, it has been estimated that within only 20 years computer memories will be as dense as the neurons in the human brain. In fact, each neuron may be equivalent to several thousand components, but even this further factor might be swallowed up by only an additional 5–10 years development at the present rate. Of course, memory capacity is *in no way* a measure of intelligence, but it is a good indication of potential 'intelligence' if only suitably clever software can be devised.

Of course, as circuitry becomes more dense, so the width of the interconnecting tracks on each integrated circuit becomes narrower. Already some circuits are being made with tracks, 'cut' by electron beams, which are only 1μ wide – that is, a hundredth the thickness of a human hair. As an idea of just how fine such tracks are, the heat from holding a circuit in ones hand would make the silicon expand by greater than the width of one track! Doubts are being raised about whether it will be practical to reduce track sizes much further in the quest for increased power and speed. Nevertheless, it is worth remembering how successful the comparatively slow 'components' in the human brain are because of the highly parallel nature of their operation. Modern high speed electronic circuitry could send signals along a computer maybe half a mile long in the time it would take a signal to cross from one side of the human brain to the others!

So what kind of hardware components will advanced computers use? Much high-powered research is being conducted into inventing substitutes for the conventional silicon-based integrated circuits currently

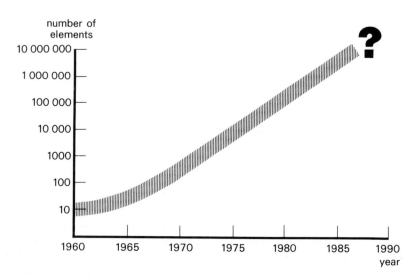

Figure 17.6 The increasing number of discrete elements on integrated circuits

used, and even for the transistor components themselves. One suggested alternative to silicon is to develop a new breed of integrated circuit based on Gallium-Arsenide (GaAs) instead. The advantages would be that there is an increase in speed by a factor of 2–3. In other words, circuits could be made 2–3 times larger (yet ‹ ₃ densely packed as before) but not drop in performance.

Meanwhile, a team at Heriot Watt University, UK, are developing an 'optical transistor' or **transphasor** which can switch beams of light on and off extremely fast. **Optical computers** built with such devices (using light rather than electricity to convey information) would potentially be able to operate at extremely high speeds. However, such devices are an unpredictably long way off. Still more esoteric research is being conducted in the field of **molecular electronics** by companies such as EMV Associates, Maryland, where attempts are being made to take the first steps towards building biological computers, in which information might be stored actually at the molecular level. Such work may indeed lead to radical new concepts in computing and robotics, but there is a very long way to go first.

Helping the disabled

Perhaps one of the most attractive of all robotic applications is to assist those who through a physical handicap are unable to live a full life. If they are lucky, the disabled have someone who will perform those necessary tasks that they are unable to do, but at best this is expensive in terms of human time and effort, and with the best will in the world it can become a mental strain for both patient and helper. As robots become more sophisticated, however, it is slowly becoming feasible (and cost effective) for disabled persons to have their own 'slave' which, under their direction, can perform at least some of the required tasks.

An example of such research is that at Johns Hopkins University, USA, involving a robotic arm and work-table system. A teachable robotic arm is coupled with a desk equipped with a typewriter, telephone, reading rack, personal computer and eating utensils. The arm can assist with such tasks as eating or page turning. Control is by means specially designed to accommodate a variety of handicaps. Similar work at the University of Tokyo involves a two-armed 'robot nurse' capable of a range of tasks, from picking up a telephone to laying a table. Another Japanese project is the 'MELDOG', intended to provide a robotic replacement to the guide dog for the blind. The mobile robot is designed to travel between selected locations using a memorised map with special landmarks, avoiding obstacles on the way. Although an interesting project, one wonders whether the appeal to the blind of a metallic electronic robot will be quite as high as that of a warm and companionable dog . . .

Laboratory work

Robots can be usefully employed in chemistry laboratories for such tasks as routine sampling and experimental work. Robots have significant advantages in terms of versatility over the dedicated sample changers on, for instance, gas chromatographs, because a robot can be programmed for use in several different routines or for carrying out a sequence of separate procedures. In addition, the microprocessor controlling the robot can also be used to accept, store and interpret data from an experiment.

Japan's Science and Technology Agency is employing a chemical robot specially designed and fabricated by Daini Seikosha to perform one step of the so called 'Maxam–Gilbert process' for cutting off a given DNA sample at a specific point. In this way the robot can alternate the tasks of analysing and identifying genes. It can also be programmed to perform about 200 other tasks, such as adding a reagent in miniscule doses of 0.0001 ml or centrifuging samples. In the UK, a research team at Manchester University have developed a commercially available 'Vuman' robot which is particularly suitable for chemical applications and manipulation or transfer of both large and small samples. Meanwhile, at the Philip Morris Research Centre, USA, a Puma robot is used to load and unload electronic balances of 0.1 mg resolution.

Marine work

Although remote controlled underwater vessels linked to the mother ship by means of an umbilical cable are quite common, there is an increasing requirement for unmanned submersibles which can operate independently of the surface ship and so, for example, move underneath oil platforms without catching the umbilical. **Teleoperation** is currently infeasible because sufficient information cannot be transmitted through water. Such **automatic robot submersibles** (ARS) are already being developed by various bodies. The University of New Hampshire, USA, has developed a rig rather like an octopus intended for complex underwater tasks such as inspection and mapping. The USA Naval Ocean Systems

Center, on the other hand, have designed a fast, long distance, shark-like device for search and inspection operations, while British Telecom, UK, have a 'Seadog' system suitable for construction, rescue and repair.

Meat processing

Although it at first seems an unlikely combination, robots are seriously being investigated for their application to meat processing. In response to a request from a meat-processing company, Imperial College, London, is researching methods for robotically deboning smoked bacon backs. At present the task is performed in the factory manually using skilled labour who cut the rib and other bones from the surface of the bacon backs using a variety of sharp knives suitable for the purpose, before the bacon is finally sliced. The job is tedious and unpleasant, and the company is finding it increasingly difficult to recruit replacement personnel for the task.

The robotic replacement for the human (figure 17.7) must first sense the location of the various bones in the given block of meat (pigs are unfortunately all different sizes and shapes!) by lowering a special sensor consisting of a series of needles over the meat. Where the needles do not sink in indicates bone. Having registered the bone layout, a special tool then starts a cut round each bone tip, and a cutting loop on

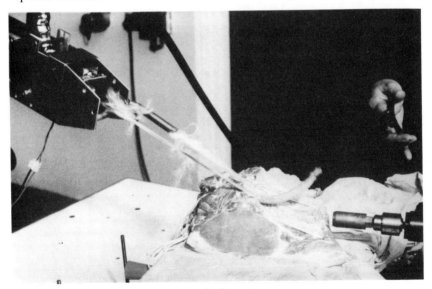

Figure 17.7 Deboning bacon backs by robot

the end of a robot arm is placed into this cut, around the bone. When the robot arm pulls back, the loop strips the bone from the meat.

Medicine

Increasingly, roboticists are seriously considering the application of robotic technology to various forms of medicine, specifically nursing care and surgery. Research has been conducted into methods of using robots to assist bed-bound patients, even to the extent of helping them into and out of bed. In the field of surgery, robots are already being evaluated for use in **microsurgery** where a highly accurate robot with substantial feedback could operate at a far higher speed and with greater precision than any human surgeon. In addition, the Tokyo Institute of Technology is developing a slender, flexible snake-like arm that will be able to probe inside patients' to perform inspection tasks, gaining access via the lower intestine. The segments of the unit are only about 15 mm in diameter, and special actuators are used based on heating special **memory metal** which then takes up a predetermined shape. It is intended to eventually use the device for administration of drugs at selected sites and for minor internal surgery.

Military work

Unfortunate though many roboticists think it is, there are several potential military applications for robotics, especially when artificial intelligence techniques become increasingly available. Robots are already being employed for handling explosives – for example at the factory run by Wallop Industries in the tiny Hampshire village of Middle Wallop. Meanwhile, the US Army Human Engineering Laboratory have developed a robotic system which can be towed on a trailer, for handling heavy shells in the field. Naturally, because this ability of robots to operate in hazardous environments extends to the battlefield itself, serious research is also being conducted into developing advanced robots which can act as soldier replacements or multipliers, and so save lives on high-risk missions.

In space, the USA announced in 1983 their intention of developing a multibillion dollar system to intercept and destroy enemy missiles. Such a system would have to draw heavily on robotic techniques, as it would require self-adaptation to new situations during periods of communication blackout. All such 'defense' research is, of course, highly funded and usually highly secret. Nevertheless, it is known that the US Defense Agency, for example, have a medium term goal of constructing a completely unmanned tank which could travel at up to 60 km per hour, avoid obstacles, locate targets and order attacks. It would be reassuring to envisage a battlefield of tomorrow (if indeed we cannot avoid one altogether) which consisted only of robot fighting robot, with not a human in sight and not a drop of blood spilt because human life was considered too valuable. The loser would be the side which first ran out of robots and realised the stupidity of sending humans in where robots had failed. It all sounds very much like a game . . .

Mining

The application of robots (or at least teleoperators) to the hazardous operation of mining has long been suggested. Recently, a 'robot drill' has been developed at Stanford University, California, which can bore through rock at angles of over 60° to the vertical, by gripping the sides of its hole, drilling out a new space and then moving into it. In this way it can drill round corners, move in a curve, drill sideways or even uphill, allowing access to locations where it is not practical to drill straight down. At first the drill is to be used for recovering oil from abandoned oil wells which cannot be reached by conventional rotary drills, but eventually it could be used for offshore drilling, coal mining, accessing geothermal wells and even for laying pipe conduits under roads. In the Soviet Union also, robots are being evaluated for rock drilling: monitoring the drilling and rock conditions in an effort to forestall drill failure due to overload.

Nuclear work

Electricity is increasingly being generated from nuclear reactor power stations, and a new generation of designs known as **breeder reactors** will soon become commercial. The environment of such reactors is so radioactively 'hot' that a maintenance man would absorb a year's acceptable dose of radiation in only one day! In order to compensate for this, it becomes necessary to employ high-sophistication teleoperators or even wholly independent robots to conduct inspection, maintenance and repair, operate in intense radiation areas and decommission plants. This last role is vital, as although it may be difficult enough to build a reactor in the first place, it is substantially more difficult to take it to pieces when its components have been contaminated by radiation.

Various research is being conducted into robotics and robotics-related technology suitable for the nuclear industry. This includes automatic decontamination during decommissioning, an underwater cleaning 'robot' developed jointly by Mitsubishi Heavy Industries and Tokyo Electric Power to clean the walls and bottom of cooling-water intake channels, and Toshiba's 2 m long robot arm for inspection. This arm consists of eight universal joints and substantial numbers of touch sensors, and can snake around inside constricted openings to inspect hidden features with a video camera.

Security guarding

Several companies, for instance Denning Systems of Washington DC, are already investigating the possibility of developing free-roving mobile robots for use as security guards. Various designs for such a security system are possible. For instance, one or many robots could patrol (randomly if necessary) constantly checking for fire, water, gas, unauthorised intruders or any indications of intruders such as lights on where there were none during the last patrol. The robot could be comparatively silent, would not require light (and so its patrol could not be viewed from outside the building), would not get bored or need to rest, it could not be bribed and it could be completely locked into its patrol area. It could regularly report to a coordinating computer elsewhere, and failure to report, or a report of something wrong, could alert the authorities.

Such robots would have the significant advantages over static alarm systems of being particularly difficult (if not impossible) to deactivate, of containing far more sophisticated sensors than it would be economic to place throughout a building, and of having the ability to check out alarm signals from static systems. If intruders were sensed then the robot could immediately photograph them for future analysis. Likewise, upon discovering a fire and alerting the authorities, the robot could immediately operate its on-board fire extinguisher or, under exceptional circumstances, be sacrificed through attempting to reach the centre of the fire before letting the extinguisher off. The less sophisticated Denning robot sentry is about a metre high and is expected to cost $25,000. The design uses sonar and infrared sensors to navigate and sense intruders.

Sheep shearing

The wool industry is in increasing competition with synthetic fibres, and in an effort to remain viable the Australian Wool Corporation has a long standing programme to seek less costly and more efficient means of harvesting its wool clip, and after substantial research has decided that the use of robots has the most promise in the medium term. Sheep shearing is an arduous task, often performed in unpleasant heat, and at remote locations, and so young people are increasingly unwilling to enter the profession anyway.

The **automated mechanical shearing** (AMS) research programme is contributed to by various sources. Melbourne University has developed a technique for automatically catching sheep, locating and restraining them in a cradle, and presenting them to the robotic shearing machine, and the University of Western Australia has devised an automatic sheep manipulator which will sequence the desired animal shearing positions. The actual shearing of a live animal requires sophisticated sensory feedback, and the University of Adelaide is developing an ultrasonic distance scanner to sense skin contours under a heavy coat of wool. Although dry wool causes few problems, wet wool can inhibit the sensor's function. Nevertheless, rising wool production costs are making automatic shearing of Australia's 135 million sheep an increasingly attractive option.

Simulation

In designing an FMS installation consisting of several cells, it very soon becomes apparent that any attempt to optimise the design manually is likely to fail. There are so many variables involved in the design that it is impossible for a human to balance them all. Somehow the relative benefits of using all the different combinations of various workpieces, tools, pallets, conveyors, vehicles, routes, operations, part types and machine groupings must be evaluated, to try to maximise potential production. For this task, computer simulation is an ideal tool, and increasingly sophisticated software packages are becoming available.

Nevertheless, the problem is a very difficult one, and any 'brute force' attempt to evaluate every possible combination is infeasible, even for the most advanced computers. Even the comparatively simple task of designing an individual cell (for instance for assembly) has far too many variables to be solvable yet by computer, although various simulation packages are available which graphically represent a proposed layout, allowing collision avoidance and other tests to be conducted. An example of such a system is 'GRASP' (Graphical Robot Applications Simulation Package), developed at the University of Nottingham, UK. This software system can assist in workplace layout, position and velocity evaluations, clash detection and coordination. An example of the graphics produced is shown in figure 17.8.

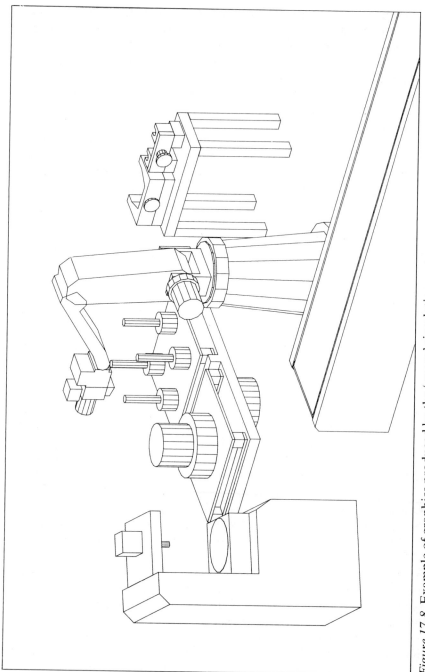

Figure 17.8 Example of graphics produced by the 'GRASP' simulation system

Computer aided engineering (CAE) systems have been around for some time. These are used for testing a proposed design without actually building it – so, for instance, a new aircraft wing design might be simulated to check that it would not fail in flight. Although such systems might possibly be useful for robot design, building a full FMS requires far more wide-ranging simulation – it is as if, when designing aircraft, we had no real idea of the optimal number, layout or detailed design of wings!

Once an FMS has actually been designed however, by whatever method, the number of variables is substantially reduced and it is at last feasible to simulate completely the operation of the system by computer. This allows the simulation to be used in the operating software to provide the online scheduling and production control necessary to omptimise minute-by-minute performance of the system. In this way, a malfunction of one FMS cell in a system can be immediately compensated for by rerouting the flow of materials to elsewhere in the factory, in such a way that the overall productivity of the new (reduced) system is as close to the original as possible.

Warehousing

Increasingly, work is being done on designing improved automated warehouses to service the requirements of automated factories. Several thousand such warehouses already exist, but research designs are becoming more and more sophisticated as truly autonomous free-roving fork-lift trucks become available, and recognition systems become more advanced. It seems likely that AGV will become particularly popular where there is medium throughput of work, and a wide variety of destinations for it. Automatic identification is usually accomplished using machine-readable codes either on the objects themselves or on **slave pallets** which carry the objects but never leave the warehouse. Overall control of the warehouse is by means of a supervisory computer which is usually responsible not only for storage and retrieval of objects but also for the ordering of stocks when they fall below a certain level.

The recent CIM trend towards 'just-in-time production' (in which inventory and **work in progress** are kept to an absolute minimum) places a great strain on warehouse control. Whereas the volume of parts stored at any one time is substantially reduced, the number of smaller transactions which need to be performed for the same production output is increased. This requires a far higher flexibility of approach than with conventional warehousing.

The incentives to introduce automatic warehouses tend to be superior customer service, reduction of excess stock by maintenance of the optimum stock level, increased efficiency resulting in reduced office and management work load, reduction in the warehouse labour force, more effective use of expensive land and improved working conditions. Among the more recent examples of joint automation of both factory and warehouse is the large automatic warehouse in the Fanuc automated factory, which extends to both floors and is used for storing unmachined, partly machined, and finished workpieces, together with components from suppliers. Parts are frequently shuttled between the warehouse and pallet stations at various machining and assembly cells using unmanned trolleys. A less conventional form of automatic 'warehousing' is employed at the library of Kanazawa Industrial University, Japan, where students in viewing booths can select any of 2,000 video and 1,000 audio tapes, which are then automatically retrieved from storage and inserted into players, by means of 'Intelibot' shuttles!

18

'Oh brave new world . . . '
FUTURE POSSIBILITIES

Prediction problems

Where are we going? Where will robotics lead us? In this final chapter of *The Robotics Revolution* we will consider what appear to be likely directions in which the field might progress. Although it is a worthwhile enterprise for any author to 'nail his colours to the mast' by predicting the future evolution of his subject, it is fraught with danger. Rarely, if ever, have such predictions actually turned out to be correct! For a start, timing is usually hopelessly out; for example, who predicted that by the 1980s every child would take a calculator into exams, yet still write his answers in long hand, on paper, using a high-technology version of the quill?

Similarly, predictions are frequently rendered meaningless by a development in some other subject – what watchmaker guessed that within a decade his mechanical skills would be made redundant because of advancements in electronics? 30 years ago, Alan Turing, the great mathematical and computer pioneer (who incidentally was the first person to program a computer to play chess), predicted that there would never be more than two or three computers in the whole of the UK, because there would not be enough mathematicians to program them! There are now several million microcomputers in the UK, many of them programmed by schoolchildren . . .

Nevertheless, attempts at long-term prediction are still worthwhile, if only because they indicate what people consider *might* happen. One person's ability to anticipate developments outside his own field (yet still of influence in his field) is probably little better than anyone else's, so the predictions demonstrate, more than anything else, the considered *potential* of the state-of-the-art of the subject; not so much what *will* happen, as what the field probably *could* develop into if it were truly independent of other disciplines.

For a subject as fundamentally multidisciplinary as robotics, the problems caused by lack of independence are magnified. Despite this, the

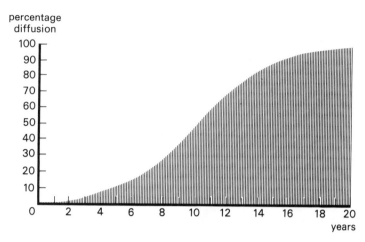

Figure 18.1 Typical rate of diffusion of even highly profitable innovations

major field which will influence it for the foreseeable future is computer hardware and software development. This is a field in which, as pointed out in the last chapter, the hardware at least has developed at a remarkably predictable rate over the last two decades. As a result, possibly the difficulty in predicting the robotic future is not so much in anticipating what may be technically possible, but in guessing how that technology will actually by applied.

Quite a lot of research has been conducted into 'diffusion of innovations', and it has been found that on the whole the spread of any innovation over time can usually be represented by an 'S-shaped curve', such as in figure 18.1. That is, at first the rate of diffusion is slow, followed by more rapid adoption which tails off as saturation is approached. This implies that the process of diffusion is rarely as rapid as might be expected. Indeed, in practice even highly profitable innovations have rarely required less than 8–10 years to move from 10–90% diffusion; the start and finish of the process have frequently added a further decade! Thus when considering the 'rapid' spread of various kinds of robotics technology, it is worth remembering that one is still referring to a comparatively gentle transition. The road from success in the research laboratory to widespread adoption in industry or elsewhere is a long one . . .

1984-1990

So, bearing this in mind and starting with something relatively easy: what about robotic developments during the 1980s? Well, it seems likely that many of the research areas covered in the last chapter will indeed graduate to being industrially feasible. SCARA-type arms seem likely to become very common for assembly work, while a few more novel designs may become established. It seems likely that many conventional designs will be 'upgraded' by using different materials for construction, such as carbon-reinforced resin. The average robot will contain far higher computing power than earlier in the decade, and increasingly textual programming (with a limited capability teach-pendant) will be used in place of purely 'walk-through' techniques. With vast numbers of people learning BASIC as their first computer language (owing to the explosion of personal computers), many robotic textual languages will remain unstructured, despite the backing of more advanced languages by companies such as IBM. Unimation's VAL could well become the *de facto* standard for this level.

As robot tasks become more complex, however, advanced languages such as AML will become increasingly necessary, and with pressure from IBM, may well become something of a standard. A new breed of 'robot programmers' (rather than personnel with other functions who nevertheless *can* program) may start to emerge. Nevertheless, developments in computing science, inspired by the fifth-generation computer project mentioned in the last chapter, will be likely eventually to simplify significantly the programming task, so making it possible for less specialised personnel to cope satisfactorily with the majority of such work.

One of the major areas likely to require such programming will be assembly. Even so, by the end of the decade, although robotic assembly will have started to make a significant contribution, it seems probable that only some of the larger companies will have actually become heavily involved, with many of the more traditional firms *still* concerned by the high-technology involved. The substantial economic advantages of using robots for assembly may by then have just started to *force* companies to robotise, but only gradually.

This process will be hastened by the availability of comparatively low-cost vision systems costing only a small fraction of the overall robot cost (as opposed to about the same). These systems will typically employ several grey levels, but colour and 3D are likely to be still only used for specialist applications. Tactile sensors (as well as simple touch) will increasingly be used, but problems in sensor design, together with an

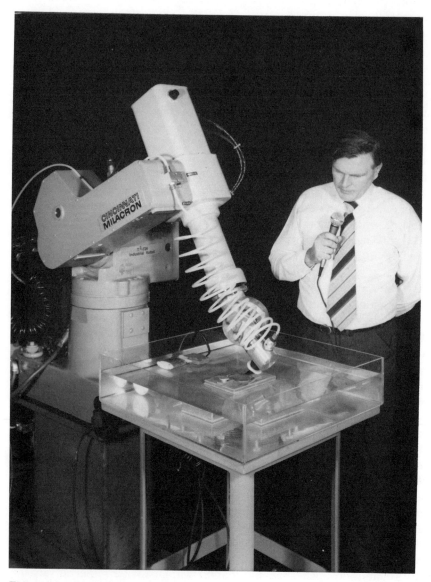

Figure 18.2 Robot programmed to engrave letters spoken to it onto a glass paperweight submerged under water

initially smaller market are likely to make them comparatively expensive compared with vision.

In addition to the processing involved with such pattern recogition, advanced computing power is likely to make itself felt in ways such as reasonable voice output as well as limited (but usable) voice input employed during programming or in emergencies. Similarly, large numbers of sensors will be able to be fitted to second-generation robots on the factory floor (although far simpler robots will still be employed for those tasks for which they remain adequate). The extra computing power available will permit far greater integration of robots into the whole factory system, together with a much higher level of automation generally.

Even so, the largely unmanned factory seems likely to remain something of a rarity while the techniques for optimising such systems are only slowly discovered and until the implications of the integration effect explained in chapter 16 become widely understood by management. A forecast by the University of Michigan, USA, states that, currently, direct labour savings represent about 60% of the total savings due to robots, yet predicts that by 1990 it will represent only about a third of the total, while inventory, space and other savings make greater impacts.

By the end of the decade, in the research laboratories, third-generation 'intelligent' robots will be just starting to appear, and the computer components in them may cost only about one thirtieth of the equivalent components five years earlier. Yet on the factory floor, this is likely to result not in cheaper controllers, but in more powerful ones. The mechanical parts of a robot arm will largely remain as expensive as before, although increased computing power may allow the inherent inadequacies of mechanically simpler designs to be satisfactorily compensated for, and inherently far less rigid arm structures to be controlled effectively.

Nevertheless, on the whole, robot-system costs (in real terms) are unlikely to have substantially altered, although for the same price the robot will have much more computing power. Similarly, its general specification will be superior, as, for example, advanced controllers permit accurate movements at far higher speeds than before. Interestingly, the University of Michigan predicts a market increase from 6–11% for robots capable of handling payloads of over 50 kg. Another market increase is forecast for very light-assembly robots. These then will probably be 'niches' in the market which will be required to be filled.

Robots will probably already be commonly used in various food-processing applications, and, outside the factory, for such tasks as routine laboratory work. It is conceivable that by 1990 robots may even be being seriously employed for such nonindustrial tasks as 'fast food' production. There is need for a word of caution here; the story goes that this has already been tried! Supposedly a US conglomerate were contemplating a completely robotised fast-food restaurant, and actually got as far as building a prototype on Long Island. The intention was that customers would select their order by pressing buttons, and robots in the kitchens would then take the necessary food from refrigerators and set it cooking. At the end of the process a single human packaged the correct food combinations, delivered by chutes in front of him. In a demonstration of the prototype the system appeared to work well. Then the human dropped a milk-shake. As he needed a replacement he reached along one of the food chutes until he found the same flavour (from a different order), and knocked over something else. Within an hour he was knee deep in rubbish . . .

1990–2000

In the early 1990s robots may start to be employed for handling flexible materials such as textiles and rubber, which requires highly sophisticated control software. Occasional use may be made of special robots in certain surgical operations, but as with undersea and space work, together with agriculture, mining and nuclear work, true robots are still unlikely to be sufficently advanced to be suitable for truly autonomous operation in such fields. Much more likely (at that stage) are forms of teleoperator devices controlled by a human rather than computer. Nevertheless, research into autonomous robots for such tasks will be highly advanced. Research into other areas however, such as robotic assistance for the aged and disabled, is unlikely to have been sufficently funded (except possibly in Japan) for there to be much evidence of such work outside the occasional research centre.

By the same period, truly robotic automated guided vehicles (ranging from simple 'tugs' to large fork-lift trucks) will have been experimentally introduced for materials transfer in selected warehouses and factories where there is a medium throughput, yet a wide variety of destinations. Nevertheless, in general, the wire-following variety of AGV will still predominate. There is certainly also the potential (and need) for robotic security guarding AGV, but whether such devices will indeed be available will largely depend on the foresight (or lack of it) exhibited by security firms.

Figure 18.3 Wire-following AGV are still likely to predominate in warehousing during the last decade of the century

Already, in 1984, a special welding robot, manufactured at the Petro-zavod shipyard, is used in Leningrad to operate in any position inside a ship's hull. Increasingly, mobility will be required for some (never all) industrial robotic tasks – especially when a robot's capabilities are required intermittently at several separate locations such as when tend-ing several devices all with long machining cycles. Suitable mobile robots will consequently be developed, together with less flexible sys-tems comprising a robot on a long 'track' – appropriate for more structured environments.

Despite increasingly sophisticated 'toys', the true domestic robot still will not have appeared in the 1990s, although many believe that it may be quite close this time – at least for some applications! Robotic vacuum-ing, polishing, table laying, loading of dishwashers and lawnmowing may all just about be *possible*, although expensive. Knowing the delight with which the majority of us seem to adore any such gimmick, it will probably not be too long before such medium-capability domestic robots become commercially available as gloriously expensive luxuries. Some simple floor-cleaning robots may even be economically desirable for use in such buildings as office blocks.

Nevertheless, the answer to the single question which roboticists must be asked more often than any other, 'Have you got a robot yet which will do my ironing and make the beds?', is still very likely to be an emphatic NO! Indeed, Kevin Dowling, a researcher at Carnegie-Mellon University, USA, has stated that a robot that does all the work in the house is at least 20 years off. Many might claim that even that was being optimistic. In addition, as Igor Aleksander of Imperial College, Lon-don, has pointed out, it is extremely likely that the need for ironing of any kind will have been eliminated by advances in fabric manufacture before anyone has produced a robot which could safely be entrusted with a steam iron and the week's laundry!

2000–2050

So much for the 1990s, but what of the turn of the century, and beyond? It is disconcerting to realise that some of today's younger roboticists will still be around in the year 2050. That is a *very* long time in robotics terms. By analogy, many people still alive today were children when Queen Victoria died. They remember life without cars and without aeroplanes, the coming of radio and then television. With one long lifetime they span the years from a time when maybe they were told that the moon was made of cheese to a time when they watched and heard men actually walking on it. One is bound to ask whether the changes modern generations will experience will not be still more incredible.

Many believe that the changes *will* be incredible, and robotics may be the greatest of all the driving forces behind them. The suggested actual forms of the changes are of course extremely speculative, but it seems likely that as computers become more and more powerful it will be natural to add sensors and actuators to many of them. The purely mechanical aspects of a robot are, after all, nothing more than a special design of computer peripheral – just like a terminal or a printer. As the computers themselves become smaller, it likewise may become convenient to house some of them actually within a mobile robot (although a computer remotely controlling a distant vehicle is also classified as a robotic system). At that stage, the conventional configuration of the 'science fiction' robot may be just around the corner.

Naturally, there is no call for sophisticated mobile robots in many applications, such as the majority of factory tasks. So, just as second-generation robots will not on the whole supplant older robots, but will instead be used for new applications, so third-, fourth- and tenth-generation robots may generally tend to all coexist, gradually creating a wide spectrum of robot types, each suitable for a different sophistication of task. However, one outcome of increased robot sophistication may be that people begin to think of second-generation robots (able to respond flexibly to changes around them) as being the first *true* robots, relegating current first-generation machines to 'programmable manipulators'.

As the twenty-first century proceeds, we are likely to begin to see large numbers of truly integrated robotic systems (factories largely) with a very large number of appendages all working in symphony. Indeed, one branch of 'robot evolution' seems likely to consist of application of robots within steadily larger and larger conglomerations of heavily integrated robotic and highly automated equipment. Such systems (not necessarily industrial or even stationary) could eventually become quite vast, with several distributed, yet communicating sections being so closely integrated that they were essentially working as one. Substantial systems might develop, for instance, for dealing with such tasks as a country's refuse collection and disposal, or, even more futuristically, assisting with municipal gardening or even policing.

It is interesting to note that a specialised form of such a highly integrated system would be one in which the system was actually self-reproducing. It was von Neumann (whose name is given to the form of architecture used in conventional computers) who suggested that self-reproducing automation would consist of an automatic mining system linked with an automatic factory which could turn the collected raw materials into a *duplicate* mining and production system.

This idea has, in fact, been extended by Georg von Tiesenhausen and Wesley Darbro of NASA, as a futuristic long-term possibility for the moon. In their plan, robot miners extract raw materials from around the factory which then refines and processes them into subassemblies. A separate production facility then uses these components to construct finished products, such as new robot miners, builders, machine tenders, machine tools and so on. In this way the overall systems would 'reproduce' at an exponential rate, not only producing more factories, but also useful 'by-products' such as human dwellings and research equipment.

Although, obviously, very far beyond the realms of current serious research, it certainly appears to many that such suggestions will indeed eventually be technically feasible, which of course in no way implies that they will actually be built. It is worth registering that such sytems actually seem to fulfil the simplistic requirements for 'life' which most people are taught at school, such as feeding, growth, movement, response to a stimulus, reproduction, 'death'. Clearly, however, most people would be extremely dubious about suggesting that such a factory *was* actually 'alive'!

In addition to such highly integrated robots (sometime during the next century?), there will, of course, be the more familiar largely autonomous robots – possibly loosely linked with an integrated system but largely independent. Although the vast majority of these robots are unlikely to be humanoid in shape, to many people it seems quite possible that, despite the prodigious advancements required in both mechanics and artificial intelligence, like the 'droids' in the 'Star Wars' films, a few of these robots will, to an extent, eventually be able to mimic humans *if so designed*. Some people argue that, 'There would be no point building a robot like a human, because in that case why not *use* a human?' However, to many this appears to miss the point – maybe it is equivalent to applying the same argument at the turn of the century to human servants and suggesting that the employers do the work themselves!

In addition, as John McCarthy, one of the founding fathers of robotics, and the man who coined the phrase 'artificial intelligence', has pointed out, there seem little doubt that a fundamentally human shape is ideal for interacting in an environment designed for humans – one only has to be in a wheelchair to realise how difficult it can otherwise be. Naturally, however, for different environments, such as inside a nuclear reactor, in space or under the sea, nonhumanoid shapes would be *much preferable*. The ideal shape for marine work may be far closer to a shark than a human! Similarly, there is no advantage in a robot being shaped like a human and being fitted with a wide range of sensors, if its sole task is to take metal ingots out of a furnace . . .

. . . sometime, never

What sort of a world might it become with so many highly sophisticated robots around? To an extent, the social aspects of this question have already been addressed in chapter 13. Yet it is interesting to compare these with two of the classic fictional predictions of the future: *1984* and *Brave New World*. Neither of these books includes any mechanical robots. There was no need – the *humans* had been conditioned into performing like robots! In this way the societies survived. Well, this book is published in 1984, and increasingly we have robots rather than robotised humans working in our factories. We have a new source of 'slave labour' which is acceptable to our modern sensibilities. But as Norbert Wiener (the originator of cybernetics) once pointed out, there is a paradox embodied in the two qualities of a slave – intelligence *and* subservience.

It is perhaps inappropriate for a book of this size to discuss too deeply the controvertial questions of artificial intelligence (AI). However, it is worth pointing out what many researchers believe may eventually be possible, while stressing that there still remain many who disagree with them! In common with the majority of his colleagues, Marvin Minsky, cofounder of MIT's AI laboratory and former president of the American Association of Artificial Intelligence, utterly believes that eventually computers will become more 'intellegent' than any human. He says, 'We just don't yet know how to weave our knowledge webs into our new machines. I see this as the most exciting research problem of our time: how to put enough mechanisms together in harmony to form minds of growing competence and breadth. Most people still think such things must be impossible to understand. I think they're only very compli-cated'.

In his textbook on artificial intelligence, Patrick Henry Winston of the Massachusets Institute of Technology points out that there are several 'myths about thinking'. For instance, some claim that computers can never be intelligent because they could not write like Shakespeare, compose like Beethoven, or think like Newton. Even if this were certain (which many would contest) the same criteria would, of course, imply that the majority of humans were also unintelligent! Similarly, those who claim that even 'intelligent' computers could only ever do what their fundamental programming dictated, forget that human learning too is initially conditioned by the programming of the genetic code.

Clearly, the implications for mankind of such 'superintelligent robots' which pass the Turing Test (explained in the last chapter), if indeed this ever happens, are inestimable. The theory goes that if humans were able to construct such machines, then potentially one or more of the machines themselves should be able to do the same, and then improve on those designs. . . Very soon there would be individual robots exhibiting intelligence greater than any human alive; then greater still. Some researchers sincerely believe that they will live to see such computers.

Winston denounces those who claim such superintelligent machines are impossible with the words, 'Of course, to believe in human superiority is a tradition. Once our earth was the centre of the universe, now it is an undistinguished planet. Once our creation was direct and divine, now some people believe it is the good luck of the primates. Once our intelligence was unchallenged, yet someday computers may laugh at us and wonder if biological information processors could be really smart. Beware of those who think it can never happen. Their ancestors hassled Galileo and ridiculed Darwin'.

Clearly, even if such experts do turn out to be correct, the form of such robots, their roles and the role of mankind, are all terribly unclear. Mankind would have to get used to sharing 'his' planet with a superior intelligence of his own (initial) conception. Of course, many would argue that to share a planet with 'superior' lifeforms is nothing new: no man is as strong as a lion or as fast as a cheetah; none can live underwater like fish or fly through the air like the humblest of birds or flying insects. It is the *type* of his particular superiority which mankind clearly values.

Yet some, including Minsky, believe that 'super robots' could even be thought of as a new evolutionary step for man himself. He argues,

> . . . how long will we tolerate the meagre years our bodies last? Our mortal stay seems fixed by makeshift engineering; our body cells, 'controlled' by programmed suicide and war, degenerate and die as immune systems fail and misinform us to destroy ourselves. I'm sick of hearing evolution praised; no self-respecting programmer would bury software bugs such dreadful ways! I'd bet we'd do at least as well to start afresh (without that billion-year accumulated mess) and try to transfer all we really want from those vast symbol-process-structure webs we call our selves into more safe and neat immortal codes.

For many, such suggestions go far too far and are oversimplistic – what, for instance, happens to 'mere humans' in such a scheme of things? Hopefully, of course, man could ensure that however intelligent robots became, they 'lived' to serve man – although some argue that it is debatable whether that would be feasible (or even justifiable) if a robot were ever, say, 1,000 times more intelligent than man. Fortunately, such problems are a *very* long way off, and some would argue that they always will be! Of far more immediate importance is perhaps whether mankind will be able to actully somehow 'link up' with robot technology as supporters of Professor Rosenbrock's approach (covered in chapter 13) would urge.

People such as Alec Robertson, a consultant on design and innovation, have suggested that responsible designers should be aiming at enhancing 'man' into 'superman', with enhanced physical and mental abilities. Robertson asks, 'Should people be kept contributing to each others lives, or should we give our responsibilities to machines where human skills and abilities may be wasted or allowed to be squandered on self-amusement in unemployment?' If the path of 'enhanced man' is followed, it is argued, then it will be possible for mankind and robots to remain on the same 'evolutionary branch', rather than man watch the robots split away.

In this way, some believe mankind will one day be able to replace its all too vulnerable bodies with more permanent mechanisms, and use the supercomputers as 'intelligence amplifiers'. That, they claim, will be the next step of participant evolution which allows man the freedom to head out into the universe, having at last escaped from the cruel fate (previously *necessary* for evolution) of being an intelligence destined to inevitable destruction through being trapped in an inadequate biological body.

This is a long way from bowl feeders, robotic spot welding, economic justification and kinematics! It can be very enjoyable to speculate about the far distant future, but now is not the time to worry about it – it may never happen. Now is maybe not even the time to prepare tentatively for it. But if one day it does come to pass, then let none say that roboticists in the 1980s did not give due warning that the Robotics Revolution was really only just beginning . . .

'I want more!'
FURTHER READING

There is a rapidly increasing quantity of reading matter being produced on the subject of robotics. Here is a very small cross-section.

General

An Introduction to Robot Technology Phillip Coiffet and Michael Chirouze – Kogan Page.
Industrial Robots: application experience H. J. Warnecke and R. D. Schraft – IFS.

Specific

Artificial Intelligence P. H. Winston – Addison-Wesley.
Automated Guided Vehicles Dr-Ing Thomas Muller – IFS.
Automatic Assembly Boothroyd, Poli, Murch – Marcel Dekker.
Numerical Control and Computer Aided Manufacture R. Pressman and J. Williams – John Wiley and Sons.
Robot Manipulators: mathematics, programmming, and control Richard Paul – MIT Press.
Robot Vision Ed. Prof Alan Pugh – IFS.

Product Literature

Industrial Robot Specifications – Kogan Page.
Industrial Robots – SME Marketing Services.
International Robotics Yearbook – Kogan Page.

Conference Proceedings

International Symposium on Industrial Robots
International Conference on Assembly Automation
International Conference on Flexible Manufacturing Systems
International Conference on Robot Vision and Sensory Control

Periodicals

Industrial Robot – IFS.
International Journal of Robotics Research – MIT Press.
Robotica – Cambridge University Press.
The FMS Magazine – IFS.

What does it all mean?
GLOSSARY OF ROBOTICS

Because of the highly multidisciplinary nature of robotics, there are many potentially confusing terms commonly employed by roboticists which are not in fact specifically associated with robots at all – such as **fettling** or **polling**. An attempt has been made to include such related terms within this glossary, together with the more exclusively robotic terminology.

absolute transducer: a transducer which outputs a unique value for any point along its travel. cf. **incremental**.

accuracy: *see* **robot accuracy.**

Ackerman steering: the form of conventional steering-linkage mechanism found in an automobile.

active compliance: compliance using sensor **feedback** to drive a robot to produce any required compensatory motions.

active illumination: technique used in vision of shining a known light source on to a scene. *See also* **structured light.**

active orienting device: device which corrects the position of wrongly oriented parts rather than just discarding them.

active transducer: a **transducer** which does not need a power source, but draws energy from the system being measured.

actuator: any device employed to cause a robot link to move.

actuator-level language: the lowest, least abstracted, level of robot programming – concerned with joint positions.

adaptive walking machine: vehicle in which conventional wheels or tracks have been replaced by 'legs'.

ADC: *see* **analog-to-digital converter.**

adhesive gripper: special design of gripper which pick objects up by 'sticking' to them.

AGV: *see* **automated guided vehicle.**

AI: *see* **artificial intelligence.**

algorithm: a sequence of computational steps for solving a given problem in a finite number of operations.

A-matrix: a **homogeneous transformation** describing the relationships between adjacent links on a robot manipulator.

AML: 'A Manufacturing Language', a **structured** robot **end-effector-level language** developed by IBM.

AMS: *see* **automated mechanial shearing.**

analog computer: a computer in which numerical values are represented directly by some physical quantity (e.g.voltage).

analog feedback: a **feedback** system which operates by the comparison of the different magnitudes of signals.

analog-to-digital converter: electronic device to transform analog signals to a digital form suitable for computing.

annealing: a process of heating and then gradually cooling a metal to remove stresses and soften it for later machining.

anthropomorphic gripper: one of a group of designs of robot grippers which to varying degrees resemble the human hand.

anthropomorphic robot: robot design (usually just of the arm) based on the joint layout of the human skeleton.

aperture time: the time required to sample and store the amplitude of an electronic analog signal.

APT: 'Automatic Programmed Tool', a programming language for describing engineering drawings, used in **numerical control.**

arc: a very bright high-temperature electric discharge caused by current jumping between two separated electrodes.

ARS: *see* **automatic robot submersible.**

artificial intelligence: the use of computers to perform operations attributed to intelligence if performed by men.

assembly surface: a component surface used in the assembly process for orientation, transport, positioning or guiding.

automaton: literally 'a self-moving machine', commonly used to refer to non-robotic mechanical moving figures.

automated guided vehicle: a mobile self-propelled and self-guiding (though not necessarily robotic) platform.

automated mechanical shearing: the technique of robotically removing fleeces from live sheep, especially in Australia.

automatic robot submersible: an unmanned underwater vessel which can operate independently of a surface-ship umbilical.

axes of motion: *see* **degrees of freedom.**

backing store: computer storage which is cheaper but slower to access than **primary store.** Same as **secondary store.**

backlash: the slight free play in mechanical transmissions before elements actually start transmitting force.

ballrace: set of trapped ball-bearings running in a groove.

ball-screw: transmission mechanism consisting of a nut on a long screw-spindle, converting rotary to linear motion.

bandwidth: the range of frequencies (and so the rate of data transmission) which a given electronic system can cope with.

bang-bang machine: an unflattering nickname for a **pick-and-place device**, due to the sound of it hitting its end-stops.

base-component structure: a form of product design in which one component acts as a base during assembly and transportation.

base system: the mathematical coordinate system (often fixed to the robot base) against which other systems are related.

basic fault: in **fault-tree analysis**, a fault which can be assigned an independent failure probability directly.

batch production: a noncontinuous production method which manufactures different products in discrete runs (batches).

binary sensing: sensing which provides information about only two possible states, such as touching/not touching.

bin-picking problem: selection by robot of an individual component from a jumbled collection stored in a bin.

bionics: creation of novel technological devices according to 'design principles' observed in biological organisms.

blow moulding: method of plastic forming where soft plastic is forced against the inside of a mould by compressed air.

blow-off system: system used on **vacuum grippers** for forcing off objects by reversing the flow of the suction mechanism.

breeder reactor: advanced design of nuclear power generator.

brown-out: a very brief interruption of electrical power which may nevertheless be sufficient to disrupt a computer.

bubble memory: a design of **nonvolatile** computer memory.

butt joint: two sheets attached by welding edge to edge.

CAD: *see* **computer-aided design.**

CADCAM: a methodology of linking CAD and CAM systems into a single integrated computerised design and manufacture setup.

CADMAT: 'computer-aided design, manufacture, and testing' – as with CADCAM but with additional automatic testing.

CAE: *see* **computer-aided engineering.**

calibration: process of moving a robot to a known position – needed when powering up robots with **incremental encoders.**

CAM: *see* **computer-aided manufacture.**

cartesian coordinates: *X-Y-Z* mathematical point convention.

cash flow: payment of actual money into or out of a company.

cast-in insert: a component made of material other than the casting alloy, placed in the mould prior to casting.

CCD: *see* **charge-coupled device.**

cell: a self-contained manufacturing unit with at least one robot and other automatic integrated manufacturing devices.

chain coding: method of recording **pixel** positions by stating in which of eight possible directions is the next pixel.

change-over system: automatic replacement of either part or all of a robot gripper to permit handling of several shapes.

charge-coupled device: electronic device in which data is stored by means of minute packets of electrical charge.

CIM: see **computer-integrated manufacture.**

circular interpolation: automatic filling in of intermediate points of a circle uniquely specified by only three points.

circular spline: component in a **harmonic drive transmission** system comprising an internally toothed solid steel ring.

closed-loop control: method where data on the actual state of a driven system is fed back to the controlling mechanism.

CNC: *see* **computer numerical control.**

commissioning: act of getting a system ready for operation.

compiler: program which 'translates' instructions written in a 'source language' into a form the computer can act upon.

compliance: limited 'give' in the rigidity of a robot arm or gripper to accommodate slight part-positioning variances.

computer-aided design: the employment of computer technology to assist in the creation or modification of a design.

computer-aided engineering: testing by computer of a proposed engineering design without actually building it.

computer-aided manufacture: employment of computers in a factory to assist with operation and control of manufacture.

computer-integrated manufacture: use of interlinked computer-based technology throughout a whole factory.

computer numerical control: use of dedicated mini- or micro-computer to implement a **numerical control** system.

connecting surface: touching assembly-component surface.

contact sensing: *see* **taction.**

continuous-path robot: robot with controller capable of specifying every point along a desired path of the arm.

continuous transfer: method of transferring **work carriers** in which each one passes the workstations at a constant speed.

core store: **non-volatile** computer store consisting of a matrix of ferrite rings able to hold magnetic charges.

CP **robot:** *see* **continuous-path robot.**

critical damping: a control method which exhibits the fastest approach to a steady state without any oscillation.

cylindrical coordinates: mathematical system which defines point positions by a radius length, an angle and a height.

cybernetics: study of the theory of control systems.

DAC: *see* **digital-to-analog converter.**

daisy chain: *see* **skip-chain.**

data base: a structured computer storage and retrieval system designed as an 'electronic filing cabinet' for data.

data capture: process of collecting raw data for subsequent use within a computer system.

dataflow architecture: design of **parallel computer** where operations on data are performed the moment it is available.

data shaping: modification of an analog signal.

DCF: *see* **discounted cash-flow.**

DCFRR:*see* **discounted cash-flow rate of return on investment.**

deburring: removal (conventionally manually) of small sharp irregularities, or burrs, from machined parts.

degeneracy: situation when two joint axes align so movement of either causes identical motion of the robot **end-effector.**

degrees of freedom: the number of distinct ways in which a robot arm can move, often corresponding to the number of joints.

Denavit-Hartenberg convention: common approach to assigning individual coordinate frames to each joint of a robot arm.

depalletisation: transfer of objects from a geometrically ordered layout to some other arrangement. cf. **palletisation**.

depreciation: notional economic allowance for the drop in potential resale value of a machine as it becomes older.

depth mapping: construction of contour maps of surroundings using depth information detected visually or otherwise.

Detroit automation: highly sophisticated **production line.**

diecasting: manufacturing process involving pumping molten metal at high pressure into moulds and letting it solidify.

die forging: method of **forging** by hammering between **dies.**

dies: the two shaped halves of a mould.

differential: form of gearing often used for robot wrists.

digital computer: computer in which data is represented by coded discrete elements corresponding to numerical values.

digital feedback: feedback using digital signals.

digital-to-analog converter: electronic device to transform digital signals (maybe from a computer) to an analog form.

digitiser: device for automatically representing actual physical locations in terms of digital values.

direct-drive robot: robot in which each joint is powered directly by a motor without any form of transmission.

direct feedback: most accurate form of feedback sensing where the actual joint state is measured directly.

direct numerical control: use of a supervisory computer to oversee the control of several NC or CNC machines.

discounted cash-flow: value of a **cashflow** which has been modified to obtain its present day equivalent value.

discounted cash-flow rate of return on investment: *see* **internal rate of return.**

discounting: modifying economic value.

discounting factor: $1/(1+r)^n$ where r is the available rate of interest and n is the number of years in the future.

distributed numerical control: *see* **direct numerical control.**

disturbance: an additional input to a cybernetic system other than that of the controller.

DNC: *see* **direct numerical control.**

DOF: *see* **degrees of freedom.**

domestic robot: a (currently largely unrealised) robot for domestic duties at the present performed by humans.

Doppler system: device measuring the change in pitch of a signal when bounced off a relatively moving object.

double gripper: robot wrist fitted with two independent grippers, the wrist being rotated to employ either gripper.

down-line loading: procedure of one computer sending a program down data lines for another to subsequently process.

downtime: proportion of its potential running time that a given robot is actually not available due to breakdown.

drift: tendency over time of a robot's mechanical response to gradually move away from the desired response.

drop forging: type of forging in which heated metal billets are repeatedly hit. *see also* **hammer forging** *and* **die forging.**

dynamic accuracy: degree to which the actual **motions** of a robot arm correspond to the desired or commanded motions.

dynamic memory: semiconductor computer store design which needs to have its contents refreshed several times a second.

dynamics: science of the action of force in producing or changing motion.

edge clustering: process in vision of joining discontinuous line segments from an **edge map** into continuous lines.

edge detection: common form of **low-level feature extraction** involving distinguishing the edges of objects in the scene.

edge map: record of the lines in an image where there is a rapid change in intensity, probably corresponding to edges.

educational robot: cheap robot not built to the necessary standards for industrial work but useful for education.

effective inertia: dynamic relationship between the torque exerted on a robot joint and its resultant acceleration.

electro-rheological fluid: liquid with the property that it solidifies when a strong electric field is placed across it.

encoder: device designed to convert angular or linear positional-state information into electrical signals.

end effector: the gripper or tool on the end of a robot arm.

end-effector-level language: medium level of abstraction programming language, concerned with **end-effector** positions.

epicyclic gears: *see* **planetary gears.**

ER fluid: *see* **electro-rheological fluid.**

error signal: signal corresponding to the difference between desired and actual states in a **servo mechanism.**

Euler angles: (pron. 'oiler') the three angles uniquely describing object orientation, as for a robot end-effector.

exoskeleton: powered device strapped round a human either for moving disabled limbs or for amplifying existing strength.

expert system: sophisticated computer system designed to incorporate the skills and knowledge of human experts.

explicit language: language which requires a programmer to specify explicitly what the robot must physically do.

external-state sensors: robot sensors providing information about the state of the robot's immediate environment.

extrusion moulding: forcing softened plastic through a heated die to produce the required product cross-section.

fail-safe: design such that any failure will in fact not directly cause danger or damage but will fail safely.

FAS: *see* **flexible assembly system.**

fault avoidance: use of highly reliable components and designs to minimise the probability of a fault occurring.

fault tolerance: use of system architectures designed to maintain correct operation even in the presence of faults.

fault-tree analysis: use of component failure-rate data plus details of system design to obtain overall failure rate.

feature discrimination: derivation of high-level image features (like area) for comparison with known object-data.

feedback: signal derived from the actual state of a driven actuator, fed back and used in the control of that actuator.

feed-forward: *see* **preview control.**

fettling: removal of unwanted pieces of metal from castings.

fibre optics: technique of passing light down a special fibre; bundles of fibres allow whole images to be sent.

fifth-generation computer: currently unrealised form of computer demonstrating substantial **artificial intelligence.**

fillet joint: one sheet joined perpendicularly to another.

fingers: those parts of robot grippers which actually grip.

first-generation robot: robots with only **internal-state sensors** (in Japan etc. refers to *pick-and-place devices*).

fitting: form of **interpolation** which finds the best 'fit' of line (or circle) on those points which are in fact provided.

fixed automation: machines without the flexibility to perform more than one task unless physically readjusted.

fixed sequence robot: *see* **pick-and-place device.**

fixity ratio: proportion of expenditure 'fixed' to a given assembly sequence compared to the overall FAS expenditure.

flash: thin piece of excess metal at the sides of a casting.

flexible assembly system: self-contained robot-based assembly system with parts feeders and conveyors.

flexible manufacturing system: robot-based system usually comprising several **cells** for manufacture of diverse products.

flexspline: component of a **harmonic drive** transmission comprising an easily deformed flexible toothed steel ring.

floor-to-floor time: time to pick up part, insert in machine tool, perform the required operations on it and return it.

fluidic logic: simulation of logical operations by varying flow and pressure of gas or liquid instead of electricty.

FMS: *see* **fiexible manufacturing system.**

forced unloading: active removal from **vacuum grippers** of light or sticky objects which might not otherwise drop off.

forging: forming metal by heating, hammering, and pressing.

form-adaptable gripper: type of **universal gripper** which somehow moulds round the shape of different kinds of object.

frame buffer: device for storing a frame of a television picture for subsequent computer processing.

frame structure: form of assembly structure in which each component is fastened onto a frame.

free transfer: form of transfer system with workstations separated by buffers, and transfer occuring only on demand.

full tracking: capability of a robot to perform a task on a moving object by adjusting its program accordingly.

functional surface: component surface used for some purpose during assembly.

gas metal arc welding: same as **metal inert gas welding.**

general sensor: highly flexible sensor (such as a television camera) capable of being employed for a wide range of tasks.

geometric fidelity: degree of distortion of the objects in a television image compared with the true shapes of the scene.

GMAW: *see* **gas metal arc welding.**

Gray code: a code in which the binary representations of consecutive numbers is only different by one binary digit.

grey levels: different distinct levels of brightness ranging from black through increasingly brighter 'greys' to white.

group technology: grouping parts based on similarity of production, so that a given group can be processed together.

hammer forging: form of **drop forging** in which heated metal billets are repeatedly hit between a hammer and anvil.

hard automation: *see* **fixed automation.**

hardwiring: linking of electronic logic components in a nonflexible manner so that they are dedicated to a single task.

harmonic drive: gearing mechanism commonly used in robot drives providing very large speed reductions in one stage.

heuristics: 'rules of thumb' used in programs which solve a problem by successively evaluating trial and error attempts.

homogeneous transformation: 4×4 **matrix** describing relative positions and orientations of two coordinate systems.

hunting: oscillation of a **servo mechanism** about the desired state with ever decreasing amplitude.

M

hybrid robot: robot design with more than one type of drive system (such as hydraulic main axes with electric wrist).

hydraulic accumulator: high pressure hydraulic fluid store to cater for flow rates higher than the pump alone provides.

hydraulic cylinder: form of linear **hydraulic actuator.**

hydraulic motor: continuous-rotation hydraulic device.

hydraulic piston: hydraulic devices which provide fixed-length linear *or rotary* movements between given points.

hydraulic ram: *see* **hydraulic cylinder.**

hysteresis: lagging behind of a physical response to the stimulus causing that response.

implicit language: programming language in which the robot makes certain decisions based on understanding of the task.

in-bowl tooling: orienting devices within a bowl feeder.

incremental transducer: form of **transducer** which measures *changes* of state rather than **absolute** values.

indexing: simultaneous movement of all **work carriers.**

indexing point: incremental transeducer reference location.

indirect feedback: less accurate form of feedback sensing which measures the movement of the servo motor not the arm.

inertial coupling: relationship between torque application at one joint and resultant accelerations of *other* joints.

inference engine: component of an **expert system** which determines logical consequences of a set of rules together.

information technology: nebulous term originally referring to data processing but expanded to include robotics etc.

injection moulding: method of plastic moulding involving ramming molten plastic in a cooled steel mould to solidify.

in-line system: transfer system layout in a straight line.

integration effect: communication increases integration of processes so allowing increased efficiency and competitiveness *overall*.

intelligent robot: *see* **third generation robot.**

interlocked enclosure: fencing with an access gate arranged so that robot operation cannot continue with a human inside.

intermittent transfer: form of transfer where work carriers start and stop at workstations for the required operations.

internal rate of return: equivalent to bank interest rate necessary to match the economic attraction of a project.

internal-state sensors: robot sensors employed for measuring the state of the robot arm itself, not the environment.

interpolation: process of filling in intermediate points.

interpreter: program which both translates and executes each user-program instruction in turn before moving to the next.

interrupt: program break caused by an external signal which temporarily causes control to be passed to another program.

interrupt handling routine: computer program invoked at once on receiving an interrupt to see what the cause of it was.

investment casting: 'lost wax' method of casting, producing very high quality products requiring little or no finishing.

IRR: *see* **internal rate of return.**

isolator: transformer or optical device designed to electrically separate two electonic devices.

IT: *see* **information technology.**

jamming: locking of one object into another, retrievable by changing direction of applied force and moment. cf. **wedging.**

jammy fluid: *see* **electro-rheological fluid.**

jointed-arm robot: design consisting of a serial arrangement of links and joints, usually similar to the human arm.

kinematic chain: treatment of a robot arm as a sequence of individual sections linked together with various joints.

kinematics: science of the consideration of movements without reference to force.

knowledge base: set of rules derived from human experts.

Lagrangian mechanics: mathematical approach involving the difference of the kinetic and potential energy of a system.

lap joint: two sheets welded together with edges overlapping

large-scale integration: electronic chip containing 100 to 5,000 discrete logic components, or 1,000 to 16,000 memory bits.

lead-screw: *see* **ball-screw.**

lead time: time from system conception to **commissioning.**

lead-through programming: form of **teaching-by-showing** where a robot is manually manipulated through the required task.

light guide: system comprising mirrors or **fibre optics.**

limited sequence robot: *see* **pick-and-place device.**

linear interpolation: interpolation of a straight line.

linear transducer: physical structure of **transducer** designed to measure **translation.**

linescan camera: camera comprising a single row of photodetectors providing a one dimensional 'slice' of a scene.

liquidity: ability of a firm to find cash at short notice to meet any business eventuality.

logical branching: ability for controller to change sequence of operations due to conditions occurring during execution.

long-term repeatability: repeatability over a long period.

low-level feature extraction: detection in vision systems of certain simple components of a scene, like **edge detection.**

LSI: *see* **large-scale integration.**

machining cell: *see* **cell.**

major axis: joint of a robot other than in the wrist.

manipulator-level language: *see* **end-effector–level language.**

manual programming: physical reconfiguration and adjustment of a **pick-and-place device** for use on a new application.

master-slave manipulator: form of **teleoperator** where a 'slave' arm mimics the movements of a manually operated arm.

matrix: mathematical array of rows and columns of quantities or symbols, convenient for advanced mathematical operations.

matrix multiplication: special multiplication for matrices.

matrix transformation: *see* **homogeneous transformation.**

mean time between failure: average time that a device will operate before any kind of failure.

mean time to repair: average time required to service and repair a device from the moment that it failed.

mechatronics: global term (wider than 'robotics') for the design strategy of linkage of mechanics with electronics.

memory metal: alloy such as titanium-nickel capable of being bent but of returning powerfully to its initial shape upon slight heating.

metal inert gas welding: most common industrial welding, using inert gas to prevent oxidation of metal being welded.

methodology for unmanned manufacture: Japanese project to instigate development of largely **unmanned factories.**

methods time measurement: system for human work analysis involving breaking tasks down into their elementary motions.

microsurgery: surgery on very small anatomical parts such as individual blood vessels or sections of the brain.

MIG welding: *see* **metal inert gas welding.**

milling machine: machining device which cuts away metal by means of a movable high speed rotating head.

minor axis: one of the joints in a robot wrist.

modular structure: structure in which various **subassemblies** are combined in different ways to produce various products.

modulation: variation of amplitude or frequency.

Moiré fringe: interference pattern appearing as dark bands caused by slightly misregistered lines, as in Moiré silk.

molecular electronics: currently impractical approach to storing data by altering the form of individual molecules.

moment: tendency of a force to rotate the body to which it is applied.

MTBF: *see* **mean time between failure.**

MTM: *see* **methods time measurement.**

MTTR: *see* **mean time to repair.**

multiple input sampling: sequentially testing several inputs rapidly enough to reconstruct the whole of each signal.

multiplexing: transmission of several signals down the same path by using rapid switching or different frequency bands.

multi-point control: term used for PTP control of robots where a very large number of discrete points can be stored.

multiprocessor system: form of computer architecture with more than one processor in simultaneous parallel operation.

MUM: *see* **methodology for unmanned manufacture.**

natural wastage: unforced reduction in workforce numbers by not replacing those changing jobs, leaving or retiring.

NC: *see* **numerical control.**

negative feedback: stable form of **feedback** where the **error signal** causes the controller to reduce the error.

neo-Luddites: hypothesised modern equivalent to the Luddites of 1811 who resisted the introduction of new technology.

net present value: economic value corresponding to the net cumulative **discounted cash flow** for a whole project.

noncontact sensing: remote sensing such as vision.

nonfunctional surface: unused assembly component surface.

nonvolatile memory: computer store which retains the data in it even after power has been turned off and on again.

NPV: *see* **net present value.**

numerical control: automatic machine control technique using prerecorded details of some or all of a machining sequence.

nut runner: device for automatically screwing on nuts.

objective-level language: highest level of **implicit language**, with AI techniques providing planning and problem solving ability.

object-level language: implicit language with high level of programming abstraction, concerned with object positionings.

offline: processor not at that time connected to the robot.

one-pass system: form of robotic arc welding which senses the seam during welding so dynamically adjusts to follow it.

online: processor in direct control of the robot.

open-loop control: potentially inaccurate control by driving actuators without then feeding back their actual response.

operating system: complex computer program designed to supervise running of other programs on the same computer.

operational research: quantification of a management problem in mathematical terms in order to find an optimal solution.

optical computer: suggested form of future computer using light instead of electricity, maybe employing **transphasors.**

OR: *see* **operational research.**

orthotics: literally 'correction of crooked parts', such as construction of robot-like **exoskeletons** for paraplegics.

overdamping: problem of system response where a **servo mechanism** only reaches a desired state after a long time.

overrun: when robot arm goes outside desired **work envelope.**

overshoot: degree to which a **servo mechanism** response to a control signal in fact goes beyond the desired value.

pallet: platform or tray for lifting and stacking objects.

palletisation: arrangement of objects into a structured geometric layout, often on a special base (or '**pallet**').

parallel computer: multiprocessor system capable of solving different parts of a single problem simultaneously.

parallel robot: configuration consisting of parallel linkages instead of the more conventional **serial** arrangement.

part-mating: process of bringing components into contact.

PAS: *see* **programmable assembly system.**

passive compliance: compliance using unpowered mechanisms.

passive illumination: utilisation only of general lighting for optical vision or triangulation systems.

passive orienting device: device which merely disgards those components which do not happen to be already oriented.

passive transducer: transducer requiring external powering.

path planning: advanced control strategy where knowledge of the whole path to be followed is used in control of the arm.

pattern recognition: area of **artificial intelligence** involved with the computerised interpretation of images.

payback: time taken for all the incomes resulting from a project to equal the original expenditure on that project.

PC: *see* **programmable (logic) controller.**

peripheral: machine which can be connected to a computer and operated under its control – a robot arm can be considered one.

permanent memory: *see* **nonvolatile memory.**

pick-and-place device: nonrobotic device which transfers items from one fixed point to another in a limited sequence.

pitch: one of three forms of object rotation, analogous to the 'pitching' motion of a boat. cf. **roll** and **yaw.**

pixel: (short for PICture ELement) one of the points which together go to make up an electronic picture of a scene.

planetary gears: central 'sun gear' with a 'planet gear' running round between it and a surrounding 'annulus'.

playback accuracy: difference of a position recorded during teaching and robot response during playback of that point.

PLC: *see* **programmable (logic) controller.**

plugboard: primitive 'programming' device where options are selected by plugging a jack into an appropriate socket.

point-to-point robot: robot for which only individual **poses** can be specified and intermediate paths are not of concern.

polar coordinates: *see* **spherical coordinates.**

polling: sequential testing of devices to see if further action must be taken, if not then the next device is polled.

port: data input/output point on digital electronic device.

pose: increasingly employed term describing jointly the position *and orientation* of a robot end-effector.

positive feedback: unstable form of **feedback** where the **error signal** causes the controller to increase the error further.

positive pressure: air pressure higher than surroundings.

predicate logic: form of logic suitable for specification of statements and questions to advanced computers. *See also* **Prolog**.

preprocessor: auxiliary computer performing some operation on data (such as simplification) to help the main processor.

press forging: method of **forging** where a hot metal billet is slowly pressed into shape between dies in one operation.

preview control: advanced control strategy taking account of path changes immediatedly ahead of the current position.

primary industries: fishing and agriculture.

primary store: principal fast memory in a computer.

principle of fixity: correlation of fraction of costs fixed to a given product and the total run length of that product.

principle of segregated cost rates: it is convenient to distinguish costs of flexible and fixed portions of an FAS.

principle of specific run-length: a given FAS is best suited to assembling products all for about the same run length.

prismatic joint: joint where the distance between it and the next varies – one of two joint types of a **kinematic chain.**

product family: products similar equipment can assemble.

production line: mass production system with a continuous stream of units sequentially built up to their final form.

programmable assembly system: same as **flexible assembly system.**

programmable feeder: feeder able to adjust its part-dependent tooling under computer control in order to accommodate changes in parts styles.

programmable (logic) controller: general-purpose micro-based limited controller which steps through a specified program.

Prolog: language based on predicate logic, suitable for AI.

proportional gripper: design of gripper capable of being opened or shut to any intermediate position by command.

proprioception: the human sense of muscular position; the equivalent for a robot employing **internal-state sensors.**

prosthesis: artificial replacement for a part of a human.

prosthetics: incorporation of **prostheses** with living humans.

protective redundancy: duplication enabling switching out of failed component, or even system, and replacing with duplicate.

proximity sensing: determination of nearby object presence.

pseudo-parallel processing: single processor switching among several systems so fast that it appears dedicated to each.

PTP robot: *see* **point-to-point robot.**

P-type joint: *see* **prismatic joint.**

puncturing gripper: gripper which pierces objects (such as pieces of cloth or sheeting) in order to pick them up.

rack-and-pinion: form of **transmission** comprising a small gear (pinion) turning against a straight toothed bar (rack).

RAPT: 'robotic APT' – example of an **object-level language**.

rare-earth materials: group of metallic elements used in construction of special magnets; e.g. samarium/cobalt magnets.

rare-earth motor: electric drive based on **rare-earth materials**, believed suitable for use in **direct-drive robots**.

RCC: *see* **remote centre compliance.**

realtime: computation performed in synchrony with the physical process with which it is associated.

rectangular coordinates: *see* **cartesian coordinates.**

refractory material: high-temperature resistant substance.

region growing: image partitioning of regions with similar brightness to correspond with surfaces in the actual scene.

relative transformation: homogeneous transformation relating the positions and orientations of two coordinate systems.

relief system: arrangement of pressure-relief valves and return pipes in a hydraulic system to prevent damage.

remote centre compliance: property allowing an unpowered mechanical linkage to act as a 'float' for a robot gripper.

remotely operated vehicle: vehicle which can be controlled remotely - if control is by a *computer* the ROV is robotic.

resistance welding: same as **spot welding.**

resolver: design of analog **rotary transducer.** cf. **synchro.**

return on investment: percentage ratio of annual profit from a project divided by capital cost.

revolute coordinates: mathematical coordinate system corresponding to a robot arm consisting of **revolute joints.**

revolute joint: joint which rotates about one axis – one of the two possible joint types in a **kinematic chain.**

robot accuracy: degree to which the actual position of a robot corresponds to the desired or commanded position.

robot repeatability: closeness of agreement of repeated movements to the same location under the same conditions.

robot time and motion: MTM-type methodology for robotics.

ROI: *see* **return on investment.**

roll: angular rotation of an object corresponding to the 'roll' movement of a ship. cf. **pitch** and **yaw.**

roll forging: imparting metal billets with a required cross-section by passing them under specially shaped rollers.

rotary system: transfer layout with a circular 'table' which brings carriers in turn to stations around its circumference.

rotary transducer: design of **transducer** capable of providing a measure of input shaft rotation.

rotary vane actuator: form of **hydraulic piston** designed to provide an output shaft rotation of less than a revolution.

ROV: *see* **remotely operated vehicle.**

RTM: *see* **robot time and motion.**

R-type joint: *see* **revolute joint.**

SCARA-type robot: rotation-rotation-translation design of arm with the major rotary joints in the horizontal plane.

scene analysis: *see* **pattern recognition.**

seam tracking: technique in advanced robotic arc-welding to dynamically follow a weld seam. cf. **one/two pass system.**

secondary industries: manufacturing, construction, mining.

secondary input: same as **disturbance.**

secondary store: see **backing store.**

second-generation robot: robot design that can significantly incorporate feedback from **external-state sensors.**

segmentation: stage in **low-level feature extraction** when images are split into areas of common brightness or colour.

sensitivity analysis: evaluation of economic uncertainty and risk by investigating a project's sensitivity to variables.

sequential joint control: simplest form of control where only one joint at a time is driven with the rest stationary.

serial-link manipulator: robot arm composed of individual sections each linked with a P-type or R-type joint.

service industries: non-**primary** or -**secondary industries.**

servo-controlled robot: robot driven by **servo mechanisms.**

servo-control valve: electrically operated hydraulic valve allowing a flow proportional to the control signal.

servo mechanism: device driven by a signal which is the difference between the commanded and actual device states.

servo-robot: same as **servo-controlled robot.**

shell mould: mould used in **investment casting** created by building up a coat of **refractory material** round a replica.

short-term repeatability: repeatability between near successive identical movements to the same position.

signal preconditioning: filtering and linearisation of an analog signal prior to passing through an ADC.

signal postconditioning: alternative to **preconditioning,** using digital-circuit equivalents *after* the ADC.

simple touch: form of **taction** involving **binary** (or continuously variable) **sensors** at only a few points.

skip-chain: ordered list of devices to be accessed and then skipped over (unless requiring servicing) during **polling.**

slave pallet: pallet which never leaves a warehouse system.

slew rate: speed at which a robot link moves once it has reached full speed after the initial acceleration phase.

smart sensor: sensor which conducts early processing of data.

software limit: constant programmed check by controller that robot position is within specified bounds and not **overrun.**

softwiring: computer based architecture – where electronic logic components are linked in ways determined by software.

specific sensor: sensor suitable only for a specific task.

spherical coordinates: mathematical system where points are specified by an azimuth and elevation angle, plus a distance.

spot welding: method of fastening metal sheets by quickly clamping between electrodes and so melting a spot of metal.

springback: deflection caused on removing an external load.

sprue: stub of excess metal formed during diecasting in the channel where metal was forced into the dies.

stacked structure: assembly component structure in which a product is built up layer upon layer in pancake fashion.

state-of-the-art: fully up to date technology.

static memory: design of computer memory which does not constantly require refreshing to maintain integrity.

stepper motor: electric motor design capable of rotating at speed or stopping at any of a defined number of positions.

stereopsis: stereo vision, as used by many animals.

stiction: (static friction) force required to initiate sliding between two contacting surfaces.

straight-line depreciation: most common **depreciation** where each year's allowance is the same throughout the project.

structural compliance: compliance exhibited by a body; the natural compliance of the mechanical linkages of a robot.

structured language: computer language with an inherent structure imposed on the logical flow through any program.

structured light: illumination used for a vision system with light projected in a particular geometrical pattern.

subassembly: partly assembled product which will not fall apart from its assembled state when moved or inverted.

surface hardening: heat treatment for metal designed to maintain overall strength, yet provide a very hard surface.

swarf: slivers of waste metal produced during machining.

synchro: design of analog rotary transducer. cf. **resolver.**

tachometer: velocity measuring device.

tactile sensing: form of **taction** using sensors consisting of arrays of continuously variable force sensors.

taction: all sensing by means of contact with an object, involving either **simple touch** or **tactile sensing.**

teach pendant: form of remote control device used for **walk-through** programming of a robot.

teaching arm: light arm with identical kinematic structure to a robot, used as a **teleoperator** during **lead-through.**

teaching-by-showing: robot programming where arm physically moves through desired motions. cf. **textual programming.**

teaching robot: *see* **educational robot.**

telechir: (literally 'distant hand') same as **teleoperator.**

telechiric device: same as **telechir.**

teleoperator: device remotely controlled by a human enabling extension of his skills to remote or hazardous locations.

tempering: form of heat treatment to reduce brittleness involving heating to medium temperatures then letting cool.

template matching: method of **pattern recognition** involving comparison of an image against existing stored patterns.

terminally-coordinated joint control: timing of individual joint motions so that they all start and stop together.

tertiary industries: *see* **service industries.**

textual programming: robot programming involving typing commands in at a data terminal. cf. **teaching-by-showing.**

thermoforming: method of plastic moulding where sheeting is heated till soft, then sucked against a cooled mould.

thermoplastic: plastic which can be substantially softened when heated, but will harden as before when allowed to cool.

third-generation robot: currently unrealised sophistication of robot displaying substantial artificial intelligence.

through-the-arc sensing: analysing **arc** current and voltage while **weaving** as a form of **seam tracking** during welding.

TIG welding: *see* **tungsten inert gas welding.**

T-matrix: product of **A-matrices** to obtain the absolute coordinates of any particular link relative to the base.

torque: turning moment of tangential force; 'twisting force'

trajectory calculation: *see* **path planning.**

transducer: device changing energy from one form to another.

transfer line: *see* **production line.**

transformation: mathematical conversion of coordinates.

translation: movement of a body without any rotation.

transmission: mechanism to convey motion from a drive unit.

transphasor: design of 'optical transistor' which can switch beams of light on and off extremely fast.

tungsten inert gas welding: variation of **MIG welding** with an electrode made of tungsten and a metal rod to fill the seam.

Turing test: hypothetical procedure based on interrogation to determine if a given computer can be called intelligent.

turnkey system: system ready for immediate use on purchase.

two-and-a-half D: (2½D) two dimensional information with some extra data on height features (though not full 3D).

two-pass system: form of **seam tracking** involving a 'trial run' with sensing followed by a welding run without sensing.

uncoordinated joint control: form of arm control where all joints are driven simultaneously but are not coordinated.

underdamping: situation as in a **servo mechanism** which continues **hunting** round desired state for prolonged period.

universal gripper: gripper design suitable for all objects.

unmanned factory: currently unrealised design of automated factory incorporating no human workers on the shop floor.

upset forging: as in enlargement of shaft-ends by hammering.

unstructured language: computer programming language with no inherent restriction on the logical flow of programming.

uptime: proportion of its potential running time that a robot is actually available for full operation.

vacuum gripper: form of gripper with one or more suction cups for attaching to the surface of an object to be lifted.

VAL: 'Variable Assembly Language', common example of an **end-effector level language**, used on Unimation robots.

variable sequence robot: true robot, capable of immediately executing different sequences merely by using new programs.

vector: quantity involving both direction and magnitude.

venturi: device capable of providing a source of suction, sometimes used as such with robot **vacuum grippers.**

very-large-scale integration: incorporation of at least 5,000 discrete logic components or 16,000 memory bits on one chip.

VLSI: *see* **very-large-scale-integration.**

volatile memory: form of computer memory which does not retain stored information when power is cut off.

Von Neumann-type architecture: conventional form of computer design, in which all instructions are executed sequentially.

walk-through programming: form of **teaching-by-showing** where a robot is remotely driven to the required point locations.

wave generator: elliptical component of a **harmonic drive** transmission acting as a rotating former for the **flexspline.**

weaving pattern: zigzagging across the general motion path.

wedging: irretrievable locking requiring parts to be pulled apart and **part-mating** attempted once again. cf. **jamming.**

Wheatstone bridge network: electrical network with four resistive sections, used in measurement of resistance.

window code: technique of avoiding output ambiguity from an **encoder** by only reading values within discrete 'windows'.

wip: *see* **work in progress.**

work carrier: platform work travels on in a transfer line.

working envelope: set of all the points correspondng to the maximum reach of the robot **end-effector** in all directions.

working space: same as **working volume.**

working volume: space enclosed by a robot **working envelope.**

work in progress: parts passing through the factory process.

world coordinate system: mathematical coordinate system referenced to the static surroundings of the robot.

world model: computerised symbolic representation of the robot arm, workspace and objects involved in a given task.

x-y table: machine surface which can be driven horizontally.

yaw: sideways rotation of an object about an axis perpendicular its top side. cf. **roll** and **pitch.**

yield: same as **internal rate of return.**

INDEX

References to figures are indicated in
bold type

torque, 55, 56, 68, 70
Toshiba, 292
track laying, 57–60, 186–9
tractors, driverless, 279
trade unions, 201, 205, 215, 228
training, 203, 219, 230, 231–2, 239,
    250, 257–8
trajectory calculations, 93
Trallfa, 167
transduction, 85, 87–92, **91**
transfer systems, 35, 94, 141–2
transformations, 47–9, 93–4, 102
transmissions, 62–3, 70, 76–9, **76, 77,
    78**
transphasors, 287
triangulation, 121, 127
Turing, Alan, 274, 297
Turing test, 273–4, 308
turn-key systems, 249

ultrasonics, 119–20, 139, 238, 276,
    279, 293
  scene analysis using, 120
UMIST, *see* University of Manchester
    Institute of Science and
    Technology
uncoordinated joint control, 92
underdamping, 87
unemployment, 199–201, 203–4, 206,
    209, 210, 228
Unimate, 18, 47, 61
Unimation, 10, 18, 19, 29–30, 39, 110,
    275, 299
United Auto Workers, 204, 217
United Kingdom, 13–14, 28, 38,
    205–6, 225
United Space Boosters, 167
United States of America, 13–14, 37,
    38, 279, 281, 284, 290
Universal Transfer Devices, 207–8
University College, London, 13, 272
University of Manchester Institute of
    Science and Technology, UK, 213
unmanned factories, 39, 277–8, **277,**
    296, 301, 305
unstructured languages, 107, **108,** 111,
    299
upset forging, 151
uptime, 241
US Airforce, 195, 283

US Army Human Engineering
    Laboratory, 290
US Defense Agency, 291
US Naval Ocean Systems Center, 288
US Navy, 193
user friendliness, 109–10, 111–12
USSR, 38, 291

VAL, 110, 299
vane motors/pumps, 63–4, **63,** 65
vendors, 38–9, 229
very large-scale integration, 33, 118
Vicarm, 31
Viking II, 194
vision, 4, 12, 104,113, 121–8, **123, 126,
    127,** 148. 161–2, 238, 299
  colour, 124, 299
  3D, 299
  2½D, 126–7
VLSI, *see* very large-scale integration
voice input/output, **300,** 301
volatile store, 95
Volkswagen, 169, 204
  Golf, 169
Vonnegut, Kurt, 200
Von Neumann, 305
Von Neumann-type architectures, 280
Von Tiesenhausen, Georg, 306
Vuman, 288

Wabot, 189, **190**
walking-beam conveyors, 142
Walter, W. Grey, 29
warehousing, 295–6, 302, **303**
Warwick University, UK, 189, 276
Waseda University, Japan, 189, 190,
    284, 285
water-jet tools, 167, **168**
Watt governor, 29, 84–5, **85, 86**
Watt, James, 34, 84
WAVE, 110
wealth distribution, 217
weaving, 93, 161
wedging, 136
welding
  arc, 46, 93, 94, 159–63, **162, 202,** 253
  joints used in, 159, **160**
  laser, by, 138, 163–4
  one-pass, 161–3, **162**
  ship, 304